SCIENTIFIC AND TECHNICAL
TRANSLATION

Isadore Pinchuck

SCIENTIFIC
AND TECHNICAL
TRANSLATION

Westview Press
Boulder, Colorado

Published 1977 in the United States of America
by Westview Press, Inc.
1898 Flatiron Court
Boulder, Colorado 80301
Frederick A. Praeger, Publisher and Editorial Director

First published 1977 in London, England, by
André Deutsch Ltd.

Library of Congress Cataloging in Publication Data

Pinchuck, Isadore.
   Scientific and technical translation.

   (The Language library)
   Bibliography: p.
   Includes indexes.
   1. Technology—Translating. 2. Science—Translating.
I. Title. II. Series.
T11.5.P56   418'.02   77-4933
ISBN 0-89158-737-3

Printed in Great Britain
by W & J Mackay Limited, Chatham

Contents

𑁦𑁦𑁦𑁦𑁦𑁦

Abbreviations

A	adjective	NP	noun phrase
Adv	adverb	O	direct object
Art	article	Oi	indirect object
Aux	auxiliary verb	*OED*	*Oxford English Dictionary*
Cn	conjunction		
COD	*Concise Oxford Dictionary*	P	predicate
comp	complement	Part	participle
D	German	pl	plural
dat	dative	Pn	pronoun
E	English	Pp	preposition
F	French	PpO	prepositional object
f	feminine	R	Russian
I	Italian	S	Spanish
Int	interjection	sing	singular
lit	literally	SL	source language
m	masculine	Sv	Swedish
MT	machine translation	TL	target language
N	noun	V	verb
Ne	Dutch	Vi	verb infinitive
n	neuter		

Symbols

🐚🐚🐚🐚🐚🐚

[ ]	enclose *phonetic symbols*
/ /	enclose *phonemic symbols*
/	between forms denotes *alteration*
:	between forms denotes *opposition*
::	sign of proportion
=	is equal to
≠	is not equal to
ø	zero
*	unacceptable form, usually a literal translation

Preface

🐙🐙🐙🐙🐙

TECHNICAL translation is today a very important enterprise, though not in business terms. Commercially it is very small fry, a cottage industry compared to the empires in other spheres of communication. But the qualitative importance of technical translation activity throughout the world far outweighs its financial significance. Dissemination of information between nations cannot take place without it. It is an integral part of modern technology, which is international and depends on the transmission of ideas across language barriers. In technology no country can afford to be an island any more. We all belong to a worldwide material culture without which we should perish.

This book is a contribution to the task of creating a better understanding of the technical efforts of our neighbours. I have confined myself to translation from German to English, because I believe that the principles outlined have some application to scientific and technical translation in general. The book is based upon a study of the theory of translation and upon practical experience in this field. The underlying viewpoint is that translation is a type of information transfer, the transfer of information expressed in one language into the terms of a second language. This outlook colours my whole approach.

I have sought to illuminate the subject of technical translating in both a negative and a positive way: negative in that I explore and, I hope, demonstrate the limitations of certain concepts and procedures; and positive in the sense that I exemplify principles that can be used to solve problems encountered in translation. I have tried to enable the translator to understand what he is doing and to give him some guidance on how to do it.

A background as a member of the staff of the Foreign Language Information Service (FLIS) of the South African Council for Scientific and Industrial Research (CSIR) has, I think, provided me with a unique vantage-point. The FLIS is primarily designed to meet the needs of research workers at the CSIR. This in itself means that it must cope with a wide and varied range of work and problems. The CSIR is an organization of scientific and technological laboratories and institutes dealing with many subject fields. In addition, the FLIS does translation work for other institutions in the country and for industry. As a result, virtually the whole spectrum of science and technology comes within its scope. Although the main languages from which translations are made are Russian, German and French, there is also a considerable demand for work in other Slavonic, Romance and Germanic languages.

This versatility is not in itself a virtue. It is not something to be sought. A translator is more efficient if he deals with only one or two foreign languages and if he is restricted to one or two subject fields. Only in this way can he become a true expert. As in other areas, specialization is desirable. But this ideal cannot necessarily be realized. Countries at the stage of development reached by South Africa must foster versatility because they have no choice. They have to make the best possible use of the expert manpower they have. This is not good for efficiency but it is splendid for the study of the theory and practice of translation.

A further advantage has been the fact that the FLIS is primarily an information service and part of a more comprehensive organization, the Information and Research Services of the CSIR. This includes the CSIR library, which is to all intents and purposes the national scientific and technical library. To have this close association with the library is of enormous value, as is the possibility of consulting the specialists in the surrounding laboratories and institutes. This helps to compensate for lack of specialization.

I should like to express my appreciation to my colleagues in our Foreign Language Information Service for their unstinting help. In particular I must mention my friend Mr Ants Kirsipuu, who combines encyclopedic knowledge with a deep understanding of the communication sciences; Dr M. D. Kranz, a veteran in information work; and Mrs M. J. Frommann, who so courageously ventured into the jungle of my manuscript and purged it of some of its worst errors. Any shortcomings are in spite of their support and not because of it.

At the beginning of 1975, I left the CSIR to become Senior

Lecturer in Translation in the Department of Linguistics and English Language at Rhodes University. The experience of conducting the first academic training course for translators in South Africa has confirmed in my mind the value of an information-oriented approach to translation.

CHAPTER I

Technical translation and information

〱〱〱〱〱〱

SCIENTIFIC and technical translation is part of the process of disseminating information on an international scale, which is indispensable for the functioning of our modern society. A technological civilization like ours is dependent for its survival on an interchange of knowledge on many levels and in many forms. This is often seen as a flow that starts with the pure scientist and ends in the products of industry. At every stage there is a transfer of information. The information need not always be in written or printed form, but this is normally the case, and the communication, storage and retrieval of information are among the dominant problems of our time. Who has not heard the expression 'information explosion'?

The three principal types of document that are relevant here may be classed as follows:

1 the results of pure science intended as a contribution to knowledge, without regard to possible practical applications
2 the results of applied scientific research carried out in order to solve a particular problem
3 the work of the technologist, which is intended to result in an industrial product or process that can be sold on the market

These three types overlap. Today's pure science may be tomorrow's technology. The discoveries of applied science and technology feed back into pure science. Faraday's experiments were prompted by mere curiosity, but they laid the foundations for the application of electrical energy in industry. What would astronomy be without the telescope, and who can imagine biology without the microscope – both products of applied science? The interrelationship is not direct and there is nothing conscious about it. The findings of pure science are not immediately applied, and some may never be. The relationship between science and technology is one of symbiosis; they do interact but they are quite distinct spheres, with separate aims and motivations.

Science can develop only through the constant creation and exchange of information. The scientist works in the glare of publicity. He seeks priority, the prestige of being the first to discover something, the acknowledgement of his peers. He wants to know what his colleagues are doing and to communicate his results to them. This interchange takes place across national boundaries.

Technology does not obey these imperatives. The demands of economic competition tend to inhibit the free and speedy transmission of results. The technologist also seeks priority, but the priority of being first in the field with the product, ahead of rival firms. The scientist, it has been said, wants to write but not to read, while the technologist wants to read but not to write.[1] In both cases the ultimate result is normally a document, and very often a document that has to be translated.

The type of document varies a great deal. In the first place there is a broad general difference between the documents that emanate from pure scientists (and, to some extent, from applied scientists) and those written by technologists. The scientist's document is most likely to be a paper in a learned journal or a paper delivered at a conference. The typical document representing the technologist's results is the patent. But between these broad categories there are many kinds of document. The channels of communication are many and varied and they cross from one language into another.

The multiplicity of documents and the diversity of languages present major problems today. The discipline concerned with facilitating the flow of communication in documents is known as information science (or information technology, as some prefer to call it, or informatics, in the terminology of the Russians). The language problem can be regarded as one aspect of the total problem of information transfer, but the study of the processes underlying translating, which is necessary if we are to reduce the problems of the language barrier, also has other features.

The proliferation of documents is very much the concern of the scientific and technical translator (who, for convenience's sake, is henceforth referred to as a 'technical translator', unless special senses of 'scientific' and 'technical' are intended). At every point in the chain from research to development and application there is an appropriate form of documentation. As we have seen, the documents vary a good deal. There is no such thing as a uniform scientific document that is used in all circumstances. Each type of document – dissertation, thesis, pamphlet, monograph, conference paper, patent, catalogue, manual, sales brochure and advertisement, to name only a few – has its own characteristics, as regards both content and language.

Translation of any one of these types of document will have its own peculiar difficulties and advantages. The translator is concerned with levels and genres as much as with working from a foreign language into his home language (i.e. the language spoken at home, which is not necessarily his mother-tongue). The foreign-language barrier is not likely to be eliminated in the foreseeable future; indeed we may well suppose that it will become more difficult to surmount. In the first place there is the constant increase in the number of publications. This automatically leads to an increase in the amount of foreign-language literature that needs to be translated. In the second place the proportion of material in foreign languages compared to that in English shows a tendency to increase. And the number of languages used for technical expression is growing. It is only in the last generation or so that Russian has been recognized as a major scientific language. In this period, too, Japanese has become important, and we are probably witnessing the emergence of Chinese. It has been suggested that English is the language of only 30 to 40 per cent of the world's publications and even that the trend will eventually lead to a figure of only 15 per cent.[2]

The demand for technical translations seems bound to increase, perhaps by leaps and bounds. This obstruction to the flow of information has long been felt and needs no emphasis. Many solutions have been offered. One is a common international language, either one of the better-known existing languages or an artificially created one. National rivalries make the former seem Utopian, while the experience of artificially created languages such as Esperanto does not encourage hope. Machine translation (MT) is of course a possibility, but the electronic computer has not yet offered a practicable answer, and may never do so. The use of the computer for data-processing – if we like to think of translating in these terms – has so far been too restricted to entitle us to expect a solution, though some applications may make the process of translating easier.

The problem is aggravated by the shortage of competent technical translators. Even if we limit ourselves to the half-dozen or so languages in which the bulk of important work is published, the problem is a very serious one. These languages are English, Russian, German and French, to which may now be added Japanese, and perhaps Italian. But in some fields, and at some levels, other languages may be important, perhaps more important than the main languages. A good deal of valuable material on coal-mining is published in Polish, for instance. Similarly, it has been pointed out that Spanish is more significant in veterinary medicine than Russian.[3] In the electro-technical field English is dominant, with Russian, German and French important, but there is also work of value being published in Polish, Czech, Hungarian and Serbo-Croat.[4]

It is instructive to look at some of the figures. It has been claimed that one third of a million or so new texts published each year are not in English,[5] and that in a number of disciplines English is not the most frequent language of publication.

A Unesco study (1957) estimated that something between 1 and 2 million scientific and technical articles, reports, patents and books were published annually.[6] A sample of 1000 periodicals showed that only 44% were in English. The rest were classified as follows: German 14%, French 13%, Spanish 5%, Italian 4%, Russian and other languages using Cyrillic characters 8%, Japanese, Portuguese, all others using Latin script and all remaining languages 1%. A more recent survey, published in 1966,[7] found that more than half the world's chemical literature was in languages other than English.

These figures, taken together with the deficiencies in the transfer of foreign-language information, help to explain the contemporary interest in a science of translation and in trying to establish such a science and develop principles of translation and aids to the translation process. This interest is not altogether new, however. The subject is one that has fascinated men for thousands of years, for we must remember that translation is an extremely ancient practice. In the Babylonian empire, for instance, it was already a well-established profession. The multi-lingual inscriptions whose decipherment has given us access to many forgotten languages of antiquity are translations. Examples are the cuneiform writings of Behistun (Old Persian, Elamite and Babylonian) produced during the reign of King Darius (522–485 BC) and the famous Rosetta Stone (196 BC) (Hieroglyphic Egyptian, Demotic Egyptian and Classical Greek). But translation must go back to times for which historical records have not been found, and indeed to eras before there was any writing. Undoubtedly men speculated on the nature of this art from the time they first practised it. In many eras in the past there have been waves of translating activity, but never on the scale of the present, and they did not resemble the modern spate of translation, which is mainly scientific and technical.

But in spite of its antiquity, translation is still a mystery in many respects today. Past writings on the subject consist of scattered insights, rules of thumb, hints on technique, speculation and some very sound general guidelines. Much of this is of value, but it is only in the last few decades that the methods of science and technology have been applied to an activity that is itself of such importance to science and technology. It is only in the last few decades that there has been a concerted, systematic and scientific attempt to unravel the mysteries of translation. This has taken place on many fronts – in connection with psychology, information theory, mathematics, anthropology and linguistics. All these disciplines have something to contribute, but linguistics undoubtedly has most to give, and translation as a discipline should be regarded as a branch of applied linguistics.

This is the view adopted in this book. It is a survey and explanation of the processes that take place in translating, which mainly uses the tools of linguistics. The aim is not to provide a recipe for instant translation – such recipes are misleading and sometimes disastrous – but to offer guidance into some of the more important intricacies of scientific and technical translating, to show what happens in the process of transfer between languages, to discuss the main problems and difficulties, the best solutions to these and the general conditions in which this activity takes place.

Scientific and technical translation is in many ways simpler to understand than literary translation. In the latter, emotive elements such as rhythm and assonance are important, whereas they play no part in technical work. We can provisionally distinguish technical writing by three main characteristics: subject matter, type of language and purpose. The subject matter is always technical, the language displays a greater frequency of technical terms than ordinary language and the purpose is always a practical one. The guiding motive of the technical text is the communication of information. It is always a means and never an end in itself, as might be said of an aesthetic work. If a technical text adorns and delights as well, this is incidental rather than deliberate; indeed it will probably be considered undesirable. The primary distinction to be made in types of document for translating is therefore that between *aesthetic* and *service* texts. Aesthetic texts include poetry, fiction, drama and *belles-lettres*. Service texts include all writings that do not come into this category. All technical texts are service texts, but not all service texts are scientific and technical. Some service texts use the techniques of persuasion or seek to arouse emotions, as with publicity material, for instance. In other fields the aim is utilitarian and objectivity is sought, yet they cannot be called scientific (except in a very broad sense of the word 'science'). One example is history. Another is jurisprudence. Technical texts such as patent literature may include references to law and may also contain historical material.

Fortunately a rigid demarcation is not necessary here. The central interest lies in texts that convey information in the natural sciences and technology. Other topics will be treated as peripheral. It is convenient to pinpoint as the subject matter of scientific and technical translation all that would be classified under 5 (pure science) and 6 (applied science) in the Dewey or the Universal Decimal Classification. This is the definition used in the 1957 Unesco report and it will be followed here, as it has the advantages of clarity and precision. At the same time, it does not entirely exclude other types of text with which the technical translator is bound to be confronted.

Service texts are therefore the broad range of texts that are mainly utilitarian in nature and commonly aim at a short-term effect. Technical texts are those service texts that are particularly concerned with the natural sciences and technology.

In the technical text, the presentation of information is the predominant aim. The other functions of language have a subordinate, even a negligible, place. The aim is above all to communicate ideas, and the expression of emotion is to be avoided as irrelevant and undesirable. In this respect the technical text is also distinct from advertising and journalism, including technical advertisements and popular science. The advertisement has very precise, utilitarian ends, and an effect in the immediate future is required, but the means used are related to the persuasive and expressive functions of language. The presentation of information may assume a subordinate role. This is especially true of consumer advertisements that use technical terms or terms that resemble technical terms, but it is also often applicable to advertising aimed at people with technical knowledge.

The most significant linguistic feature of the technical text is its vocabulary, the specialized terminology of the particular discipline. The Russian writer Fëdorov considers technical terminology and technical phraseology to be the distinguishing traits of a class of translation literature that comprises news (i.e. journalism), documentary material (commercial and official) and science. [8] While we might agree that these types of text have a good deal of common vocabulary, the vocabulary is not used in the same way. The difference in use of vocabulary is a very important element, since the meaning of words is largely determined by the context in which they are used, but I shall say more about this later.

The grammar of technical writing does not differ very sharply from that of other kinds of writing. They all share the common grammar of the language, though there are some striking peculiarities and tendencies in the morphology and syntax of technical prose. The great frequency in the use of the passive in technical English comes readily to mind. In many ways the features of technical language are also those of officialese – it is a highly stylized language and would sound stilted and pompous if used in everyday speech. Some writers consider this stylized technical prose to be barbarous and advocate a more straightforward mode but, as we shall see, this is not quite as simple as it appears. The normal rules of clarity of expression do not always apply to technical texts. Thus the high frequency of the passive, which some have condemned, serves the informative purpose of technical writing and the suppression of emotion and individual personality.

Although technical writing tends to be highly stylized, it should be stressed once again that technical prose, for all this, is not essentially a different language from our ordinary everyday language. It is at most a special variety of formal language, and one that is expressed almost entirely in written form. It is part of the general language. Its grammar and vocabulary are drawn from the general language, but sections of its vocabulary acquire specialized uses and other sections are more or less confined to specialized fields. Hence what is said about technical texts is also a statement about language in general. Observations on technical translation will be relevant to translation *per se*. If it is true that not every problem of translation is relevant to the technical translator (transferring the effects of rhythm, metre and so on), much of our concern does have a wider application than to technical translation alone.

The advantage of technical translation from the translator's point of view is that it is undoubtedly more restricted in range than aesthetic translation, but its simplicity can be exaggerated. It has been thought of as a kind of engineering operation in which standardized components are fitted together, or as a similar process to the solving of a jigsaw puzzle. But these analogies are very misleading.

At the other extreme it has been said that translation, and technical translation is included in this judgement, is not possible at all. Ortega uses the example of 'set theory' in mathematics, which its creator, Cantor, called *Mengenlehre* in the original German. In Spanish it becomes the *teoría de los conjuntos*. *Menge* (literally: quantity, number, amount, crowd, mass) does not have the same range of meanings as *conjunto* (aggregate, whole), writes Ortega,[9] and he warns against exaggerating the translatability of the mathematical and physical sciences. I take his point and will discuss it in due course. But even Ortega admits that natural-science texts are more translatable than those in other fields.

The example of *Mengenlehre* in fact exemplifies a major problem. It can be viewed in terms of a difference in total outlook (world view) between one speech community and another. More simply, we may draw the conclusion that translation equivalents between languages are generally only approximate. They do not correspond completely in their range of meaning. Sometimes this does not matter much, but on occasions the value of a whole text may hinge on one expression such as this.

There are also a number of problems of a more practical nature. When Ortega speaks of the 'misery and splendour' of translation, he is thinking mainly of literary work. Technical translation would seem on the face of it to have more misery than splendour about it. It is not as exciting as literary translation, which is a creative, artistic activity. It does not offer the most glamorous of careers. This is one reason why there is a shortage of properly qualified technical translators. Another reason is the incorrect conception of translating prevalent among the general public, which tends to lower the status of technical translators. In fact technical translating demands high qualifications if it is to be done properly. While it is no more to be compared with aesthetic translation than the writing of a scientific paper is to be compared with writing a poem, it does have a creative element; it demands intelligence, ingenuity and a great deal of knowledge. While it involves much routine work, it does have its moments of 'splendour', Since it is primarily concerned with the transfer of information, it goes without saying that any specialized knowledge is a great advantage for translating – in that field, and perhaps in related fields. Naturally one must also have a good knowledge of one's own language and the foreign language. There is also such a thing as a flair for translating, though this is hard to define. The relatively high proportion of standardized terms in technical texts is a great advantage to the technical translator. Against this we must set off the inadequacy of other aids, particularly technical dictionaries.

The practical and the theoretical elements are very closely associated in translation. Understanding of language structure and specific problems in the transfer of ideas between languages will help in practice. The insights derived from practice help in this understanding, creating a beneficent circle. Few branches of linguistics – and as we have seen, technical translation should certainly be seen as a branch of applied linguistics – show such close interrelation of the practical and theoretical aspects.

CHAPTER 2

The nature of language

〣〣〣〣〣〣

THE most important characteristic of language for the translator is that it is a system of signs used for communication between human beings in society. This is not a complete and perfect definition of language but it offers an adequate explanation of the linguistic processes underlying translation.

System implies that the signs are interrelated within a totality; *social communication* that it is not individual expression as such that concerns us; and *sign* conveys the information that the elements of language are conventional. The sign nature of language means that there is nothing inherent in, say, a word that makes it the appropriate name for an object. We call a certain animal a *dog* not because 'it is a dog', not because there is a mystic rightness in so doing, but simply because we have been taught to do so. There is, so to speak, an implicit agreement among English-speaking people that a certain animal should be called a *dog*, just as there is an agreement among German-speaking people that it should be called *Hund* or among Frenchmen to call it *chien*. The sign is not inextricably attached to the thing but tied to it only by convention; in this sense it is arbitrary. Meaning shift does take place and new signs become associated with objects. The Anglo-Saxons used the same expression as the Germans for 'dog', *Hund*, but at some stage after the Norman Conquest this was replaced by *dog* and the word *hound* acquired the specialized meaning of a dog used for hunting. Thus the connection between thing and name, or rather between concept and linguistic element, is an arbitrary one.

Every language is a system of signs. A given language functions as a whole with all its elements interrelated and influencing one another. A change in any part of the system will affect the rest. The effect is greater between some parts of the system than between others – these are more closely related groups of signs. An example is the range of colour terms that forms a sub-system

22

within the main system. The signs also combine with one another in a fairly regular manner and according to rules. Some of these combinations link very small units like affixes and words, others link words with other words (compounding) to form larger word units, and still others are loose combinations of words to create phrases and sentences. These elements form hierarchies. A word may be part of a phrase but the reverse is impossible; similarly phrases can combine to form sentences but not vice versa. Each language system has its own specific rules for the order and manner in which these combinations occur.

The language system is acquired by each individual born into a particular community; it is learnt in the normal way from infancy onwards. It is handed down in this manner from generation to generation. Each individual draws on the resources of his language system and uses it in accordance with the needs of the situation and his own degree of mastery of the language. This contrast between the language system as such, which ordinarily we call the German, the English, or the French language, and the individual's handling of it is very important. Every text that is translated represents an individual usage of a language system and the translator seeks to translate this individual usage into the terms of another language system. This is so even with technical texts, although in technical writing the aim is generally to reduce the individual element to the minimum.

Since each language is a peculiar system that can be distinguished from other systems of the same kind, a language system is sometimes spoken of as a *code*. This is a useful term as it draws attention to the structural nature of the language, the fact that it exhibits regularities and that certain forms and types of behaviour are repeated. As a code, language can be conceived of as a list of items and a set of rules for combining those items and producing modifications in them. This is very convenient as far as it goes. But the word *code* is also used in a narrow sense. When it is used as an equivalent for language system, we say that English is one code and that German is another. The process of translating can be thought of in terms of *code-switching*. But where the word code is used to suggest that language is an artefact, an instrument deliberately made by man for a specific purpose, it is misleading. It is then on a par with the Morse Code and the commercial codes which are infinitely simpler than any natural language system. In fact these codes (and more complex ones) are secondary systems that have been derived from natural language. The Morse Code is merely a way of representing the alphabetical signs of ordinary language. Commercial codes are based on the high frequency of certain stereotyped phrases within certain contexts. All codes in this sense are more restricted than ordinary language. This also applies to the use of code to denote language variety, although here we are dealing with natural language. A language variety may be in a very formal mode as compared to a colloquial style of expression. Changing from the one mode to the other is also called code-switching, and in fact this is the more usual meaning of the expression. It involves switching from one code to another *within* a language system. The behaviour of bilingual people, who make rapid transitions from one language to another, entails code-switching between language systems and has many interesting parallels to the process of translation. In both processes there is a tendency for one language to interfere with the other, that is, for the patterns and items of the one language to be used (wrongly) in the other. It is important not to confuse the derived codes, which are finite and closed systems, devised for specific purposes, or, at best, limited systems used for special purposes, with the natural languages, which are infinite and open systems. In this sense language is like a biological organism, but it is also very much associated with human consciousness.

A given language is a structured system used mainly for communication in society. It is an instrument only in the sense that human society is an instrument. It is in this sense an 'instrument' at the disposal of individuals who wish to communicate with one another, but it has an existence over and above the individual. It has historical existence. It is 'inherited' or, as we have said, learnt by the individual. It has some of the characteristics of an institution in that it is greater than the individual, but it also resembles an organism because it displays growth and change. It finds expression only in individual usage.

The items that we associate most commonly with language are words. We tend to think of language as composed of words and we take it for granted that we know what a word is. Generally speaking, of course, we manage perfectly well, but the translator needs a somewhat more sophisticated definition. The first thing to be noted is that the word is not a picture of anything, though *some* words in a language do become associated with objects – *horse*, *tree*, what you will. The situation is clearly less simple when a word is associated with something other than a concrete object, a quality like *warm*, say, or a concept like *beauty*. But even when we have a word associated with a concrete object, it is not a simple case of one word standing for one object. A word represents a complex of features. Every item has more than one feature by which it can be named. The word *water* is understood as referring to a well-known substance occurring in nature. We can think of it in terms of its fluidity ('flowing like water', as in 'money flows like water') or its odourlessness and tastelessness when it represents purity, or used to dilute something else, and so on. When we use the word 'water' in a sentence, we are selecting any one of a number of possibilities associated with it: 'He gave him a glass of water'; 'He watered down his argument'; 'He watered his horse', 'Blood is thicker than water' and so on.

At the same time a number of other words have meanings close to *water*. Each shares some of the meaning of water but also has some other meaning. For example, *moisture* may be due to water and usually refers to a form of water, but it may be moisture from other fluids. *Dew*, again, is normally a specialized form of water. It shares with *moisture* the idea of a small quantity, but occurs in specific circumstances. *Dew* always means moisture too, but the reverse is not true. *Moisture* is therefore a more generalized word than *dew* but a more specialized one than water. In the *Concise Oxford Dictionary* dew is defined as atmospheric vapour 'condensed in small drops on cool surfaces' and is contrasted with another specialized word, *rain*: 'condensed moisture of atmosphere falling visibly in separate drops'. The mere existence of these related words affects the use of the item *water* in the language. In a language which had no special word for *rain*, for example, we might find sentences like: 'Water fell from the sky in separate drops'. Because the word *rain* exists in English, it would be odd to talk of 'moisture dropping from the sky', or 'water dropping from the sky' instead of 'rain falling'. It would be even less accurate to speak of 'moisture in a flask' when what was meant was 'water in a flask', even though the moisture might be water.

A language element such as a word functions in two ways. It refers to something else (*water* points to a fluid substance with certain properties), but it is also a member of a language system and exists in relation to other language elements. Thus we could say that the meaning of the word *water* in English is that it denotes H_2O and that it is related to other words like *rain, moisture, dew*. Other relationships and contrasts can be mentioned, such as liquid as against solid, water as against other liquids (oil), as something to drink, and so on. For the moment, however, let us consider only this primary distinction between the word as indicating something outside language and the word as related to other words within the language.

Its relevance for the transfer between two languages can be illustrated by the example of *sheep* and *mutton*. In English *sheep* usually designates an animal and *mutton* the meat obtained from the animal. But in French *mouton* does duty for both. In translating from English to French that may create no difficulty but proceeding in the reverse direction may be troublesome. *Mouton* has two translation equivalents in English, either *mutton* or *sheep*. It designates in the French language system what the two English words together designate in the English language system. It is convenient to say that *mouton* has the same *meaning* as *mutton* and *sheep*, but its value within the French language system is a different one from the value of *mutton* or *sheep* within the English language system. Each of the English items has a more restricted range of application than the French. Alternatively, each of the English items can be said to be more specialized than the French one. The correct English translation of *mouton* depends on the context of the French sentence in which it appears, and perhaps on more information than the context provides.

By value I mean the role of the item within the system, which is dependent on its relationship to other items. Meaning, as I use it here, is the relationship of the item to the outside world that is created and maintained by usage or convention. The item does not directly represent an object in the outside world.

A word may relate to more than one object in the outside world. The expression *nut* may refer to a fruit, to a piece of metal, to an eccentric person and to a number of other things. Conversely, more than one word may represent a single idea, e.g. The Royal Society for the Prevention of Cruelty to Animals; the White House. There is no one-to-one relationship between word and thing (or idea).

Obviously it is less likely that the value of an expression within its language system *and* its meaning will correspond exactly with the value of an expression within another language system *and* its meaning. It is more likely that meaning will overlap, as with the *mouton*: *sheep* example, and that there will be a divergence in value. Usually there is considerable disparity in the values. It is this divergence between value and meaning that gives rise to many linguistic howlers. The disparity comes to light among other things in the kinds of application possible for a word in an SL (*source language*: the language from which the translation is being made) and that possible for its counterpart in a TL (*target language*: the language into which the translation is made). There may be a wider or narrower range of application in the SL than in the TL; or no application may be possible in the TL, though the words may be equivalent in terms of meaning.

The phenomenon observed with *mouton* is also found, for example, with the German expression *Fleisch*. This designates the same thing as *meat* or *flesh*, but its range of application, as with *mouton*, is wider within the German language. It can do within the German system whatever the two English items can do within their system. But *meat* cannot function in English in a manner exactly parallel to *Fleisch* in German. This is also true of *flesh*. We can for instance talk of a 'flesh wound', but never of a 'meat wound'. *Flesh*, of course, has a wider range than *meat* and can be used in contexts similar to meat. We can talk of 'roast flesh' in the same context as *roast meat*, but there is always the possibility of this referring to an accident in a fire in which a human being has been burned. We can talk of 'the flesh of an animal' intended for eating, but whereas it is in order to talk of *flesh-coloured stockings*, it would not do to say *meat-coloured stockings*.

Thus the value of *Fleisch* within the German language system is not identical with the value of either *flesh* or *meat* in English. It refers to the same thing, but its relationship with other linguistic items is different.

This can be illustrated in another way. If we use the expression *Herr Schmidt* in a German sentence, the effect will be different from that within an English sentence such as: 'We expect a visit from Herr Schmidt tomorrow.' This would also be different from: 'We expect a visit from Mr Schmidt tomorrow.' The use of 'Herr' in the English sentence gives it an emphasis greater than the use of 'Mr', since it stresses the Germanness of Herr Schmidt. In German, of course, it has no special emphasis of this kind. It acquires a value within the English sentence that is different from the German because it contrasts with English expressions like 'Mr'. It has, among other things, the value of 'foreignness'.

Phenomena of this type are of great importance in translation and are sometimes used as an argument to show that translation is not really possible, since this can take place only when value and designation coincide, as in *zwei* and *two*. We have said that neither *meat* nor *flesh* fully translates *Fleisch*. However, there are many ways of compensating for this. Naturally there are also occasions when no compensation is possible and the expression simply cannot be translated.

The term 'value' is used with reference to the vocabulary of a language, the inventory of its lexical items. This inventory is potentially unlimited. It is constantly undergoing change in a number of ways. This, of course, is a complication for the translator. Another is the sheer size of the vocabulary, which no individual can hope to master, and which in fact cannot really be contained in a dictionary. The available vocabulary of a language like German has been estimated to be between 300,000 and 500,000 words.[1] The lexicon of a language is indeed formidable.

Grammar is easier in this respect – there are fewer items to consider. When we speak of the grammatical structure of a language, we take account of the fact that the items of the language combine with one another and occur together in accordance with rules peculiar to that language. These rules affect the type of combination and also its sequence. Combinations can be made between the smallest individual units of vocabulary like *morphemes*, which occur, for example, in affixation and derivation, or units of common usage like *words* in the form of compounds, phrases, clauses and sentences. The rules of sequence are those that make it normal for an adjective in English to precede the noun it qualifies and for a certain order of words to be used in clauses and sentences (syntactic rules).

Each language has its own rules for these events. It has, so to speak, its favourite patterns. Some patterns may coincide in different languages and these can simplify translation. Languages which have a common ancestry or which have been strongly influenced by one another are likely to display such overlapping. The Romance languages descended from Latin, for example, have a similar method of compounding, which differs from that found in either German or English, which are also of common origin. On the other hand, many Romance forms are found in English, so that it has sometimes been called, half-seriously, another Romance language.

The genetic classification of languages, however, is not as important as a typological classification. It is very relevant that English is, for example, what is called an isolating or analytic language – one in which the words are invariable – and that German is of the inflecting type – the nouns are declined, the verbs are conjugated and so on. This is a relative classification, of course, since English is not absolutely analytic, as it does exhibit some inflection (note, for example, the pronoun system). Nor is German absolutely synthetic. But compared with English, German is certainly a synthetic language. The translator becomes very much aware of the structural differences between the two languages. Spanish and Italian exhibit more inflection, specially in their verb systems, than English, but in noun declension and certain other features they are more analytic than German.

The fact that languages display a hierarchical organization is of importance here. In a given sentence, certain elements will be dependent on others. The primary elements will be the subject and the predicate. In the subject, the main element is usually a noun and it is accompanied by subsidiary elements such as article, adjective, adjectival phrase and so on. The main element of the predicate is a verb, with possible subsidiary elements such as various complements and adjuncts. Translating entails an analysis (not necessarily a fully conscious one) of an utterance in this manner. It entails other types of analysis as well. But it undoubtedly includes a grammatical analysis. We pull out the main elements from the sentence and their subsidiary elements, noting the various relationships of dependence, in order to arrive at the meaning. This is normally performed intuitively. A process that is similarly intuitive reverses the analysis and the utterance is reconstituted in terms of the TL. All these processes will seem to be taking place simultaneously and it is only when difficulties arise that the translator becomes aware that he is carrying out operations of this kind. It is in fact very useful to be aware of the nature of language in this way in order to cope with problems in translating.

In regard to this hierarchical structure, it is possible to speak of grammatical levels of ascending importance: morpheme, word, phrase, clause and sentence.

CHAPTER 3

What is translation?

🕲🕲🕲🕲🕲

THE nature of translation is a topic that tends to lead to extreme points of view. There is a tendency on the one hand to emphasize the role of language to the exclusion of everything else, and on the other to neglect the role of linguistic elements and to concentrate on the conceptual content. The notion that translation involves merely replacing words in one language with words in another is probably the most common one held by the general public.

But if translation were solely the replacement of words, the appropriate procedure would be to consult a bilingual dictionary. This would contain a list of SL words with a corresponding list of TL words; each SL word would have its TL counterpart. This indeed is the naïve idea of a bilingual dictionary and what the man in the street expects to find in it. It presupposes that words are clear-cut and distinct entities, each word normally having only one clear and distinct meaning. This idea goes with an equally widespread notion that language itself, any individual language, is a list of words of this kind and that each word is the name of some object in the external world. This has a precedent in the account of naming in the Book of Genesis, and underlies the view that translation can be simple word-for-word substitution.

Anyone who has ever attempted translation will be aware of how illusory all this is. Yet it persists, and in a somewhat more sophisticated shape provided the starting-point for the first experiments in translating by means of the electronic digital computer (machine translation). Part of the reason for its persistence is that there is an element of truth in the idea that translation is concerned with words only. Translation *is* concerned with words, but not with words alone. The mechanical substitution of word units can be effective under certain conditions and up to a point. Some simple forms may be converted in this way from one language to another, A phrase like German *der Vater* can be replaced in

English by *the father*. We can arrive at this with an ordinary bilingual dictionary, and there are a number of other phrases that can be dealt with on these lines. For example, a phrase containing an adjective, like German *der gute Vater*, will with a little extra information yield English *the good father*. The information that the adjective *gut* takes the ending *-e* before the noun *Vater* may be contained in the dictionary. With a structure rather more complicated than this, dictionary translation of this type tends to become very difficult, if not impractical, even if the process itself is not very much more complicated. To arrive at *the father goes* from *der Vater geht* requires information about the behaviour of the verb *gehen* that is more readily found in the grammar book than in the dictionary.

Even before we reach this stage, and when we are still dealing with isolated words and simple phrases, certain phenomena strain the resources of the dictionary. A number of words in the language can scarcely be called the 'names' of things, whatever may be argued about words like 'horse' or 'dog' (and even these, as has been noted in the previous chapter, are extremely complicated). It certainly cannot be said that prepositions, an important class of words in English, are the names of anything. What do prepositions like *on, to, auf, zu* name? If it is claimed that they name ideas, such as ideas of spatial position, it would be difficult to demonstrate this clearly. We say *on top, on the mark, on the whole, carry on, auf Erden, auf der Strasse, auf dem Markte, den Blick auf einen richten*, and *zu Anfang, zum Beispiel, zu Berlin, go to town, he wants to go, to the fore*, and so on *ad infinitum*. What do all the *on*s, the *auf*s, the *zu*s and the *to*s have in common? What spatial position or direction do they signify? On the contrary, each example signifies a different idea. In *auf Erden* the idea is 'on' (on earth), but *auf der Strasse* carries the notion of 'in' (in the street'). With *auf dem Markte*, we are *at* the market, but when we say den *Blick auf etwas richten* we signify that the gaze is directed *towards* something.

The meaning of these prepositions varies according to the context, but this being so, they cannot be 'names' in the sense that one could say that 'Tom' is a man's name or that 'dog' is the name of a particular kind of animal. They are items with a very vague meaning that is made more precise by the context. A dictionary may give a whole list of these uses, but naturally this must be limited.

The dictionary can also inform us of the different meanings of words that stand for more than one thing – like *crane*, which can represent either a bird or a lifting device, German *Schloss* (castle, lock or bolt), or *Mark* (unit of money, boundary or marrow). But it cannot tell us how to select the right meaning when the word occurs in a text for translation. Words, as we know, rarely occur in isolation, but in combination with other words. As with *auf*, *on* and *to*, *Schloss* or *Mark* can usually be interpreted by means of their context, but there are instances where context is not enough. These examples should warn us that word-for-word substitution is unrealistic as a general method of translation, but they do not necessarily show that translation is not a matter of words. What they do show is that account must be taken of the rule that a word is modified by the company it keeps.

However, so far we have considered only words and phrases. The limits of simple word-for-word substitution become even more obvious when we try to translate sentences. To return to *der Vater geht*, unless there were additional information, the result of a simple replacement process would tend to be *the father go* – a kind of pidgin English. The additional information is of course grammatical. To make a correct English version, we must be aware that *geht* is a special form of the verb *gehen*, that it has the notion of a third person (he, she or it performing an action), that it contains the notion of singular (only one person acting as against more than one) and that it contains the notion of the present tense (something happening currently). This knowledge enables the translator to make the transformation of *go* into the correct English verbal form of third person, singular, present tense, which is *goes*. Extra grammatical knowledge, elementary knowledge at least, is necessary if we are to be able to go beyond words and phrases and translate sentences. The sort of sentences that we can translate in these terms is another matter. The example given is a very simple sentence form and only sentences modelled on it can be translated in this way. There are many complications which can be introduced, for which greater grammatical knowledge will be required. A 'rough translation' can be obtained that may seem adequate under some circumstances. But the reader will have to expend a great deal of time and energy to decipher it and it may also contain many errors.

Included in this extra knowledge, even for simple translations, must be rules of word order. In a language like English, word order determines meaning and certain sequences are obligatory. Thus *Jack strikes Pete* has a completely different meaning from *Pete strikes Jack*, but a reversal like this in German may not have the same effect. *Der Sohn liebt den Vater* has the same meaning as *Den Vater liebt der Sohn.* (The latter word order is, of course, not in common use.) Each language has its own grammatical rules. Translating from German to English, we must bear in mind that in German the subject of the sentence need not precede the verb, since its role can be indicated by other means than word order. In English the order is all-important, and the subject must normally come before the predicate verb. In translating from a Romance language into English we must obey the rule that in English the adjective comes before the noun it modifies. In Italian *casa bianca*, as in Spanish *casa blanca*, the adjective follows the noun. The literal translation **house white* would be ungrammatical in English; we must say *white house*. Similarly, the article must come before the noun in English (as in German, French, Italian and Spanish), so we have to write *the crane* and not **crane the* if we wish to be intelligible; but in Swedish we say *kranen*, with the article -*en* at the end.

Further problems arise from the fact that combinations of words do not necessarily represent straightforward additions. In language, $1 + 1$ does not always equal 2. *Blackbird* is not the additive result of the equation *black + bird* but a new entity: 'bird' certainly, but not the same as 'black bird', any more than a 'blacksmith' has to be a smith who is black.

The opposite pole to translation seen as a 'word-bound' phenomenon is a 'thing-bound' activity in which the role of words is purely instrumental. The transfer of concepts is the be-all and end-all of the process. Therefore knowledge of concepts is all-important; linguistic knowledge counts for very little. This idea is especially prevalent in technical translation, where understanding of the subject matter is obviously of the highest importance.

This does not imply that concepts can be conveyed without language – this is an entirely different matter – but that language is purely and simply an instrument, nothing more than a channel through which the concepts flow, and not in itself presenting problems, except that it may be a clumsy tool.

The form in which this view most commonly appears is the suggestion that a combination of word-for-word substitution aided by the dictionary plus subject knowledge will suffice to produce a satisfactory translation. It is a procedure which has been recommended, for example, in the translation of chemistry texts, where many internationally accepted symbols are used and where, to some degree, an international nomenclature is found. A trained chemist, it is argued, can find his way through a text in a foreign language with no knowledge of that language, or with minimal knowledge. It is difficult to determine the extent to which this is true. Certainly it does not apply to all texts, and there may even be instances where subject knowledge plus lack of linguistic knowledge may be dangerous. The chemist may, for example, discover erratic reasoning and correct it in the belief that his is the correct version of what the original author said. But the original author may have been wrong and the translation should have preserved the error! It should be recognized that the dependence on subject knowledge alone may not be adequate, even in highly favourable cases like chemistry (or mathematics).

The formula for this kind of translation is pidgin + extra-linguistic knowledge = satisfactory version. The formula for word-for-word transfer can be stated as $W' = W''$. This formula assumes that the information is encapsulated in some fashion in the words, and that if we attend solely to the words, the information will automatically be transferred. Look after the words and the concepts will look after themselves. This conception tallies very well with a code-switching notion of translation. The text is an expression of a compact system (code). The chain of communication is something like the following: the author encodes a message in the SL; this coded message is then decoded by the translator and re-encoded into the TL. With an experienced translator, and especially between languages that have been frequently translated into one another, the process may seem like a fairly rapid and smooth switching from the one to the other.

The concept-for-concept translation ($C' = C''$) attempts to acquire the information in a more direct manner, to take it, as it were, by storm. Our chemist will recognize the equations and formulae in a text written in Japanese with a script that is not even alphabetic, and, with a few clues about the language, he may manage to decipher some of it. If technical texts were expressed entirely in a symbolic notation, the formula $C' = C''$ would be perfectly practicable; and if every item of vocabulary in the text expressed one concept and one only, technical translation at least would be one of the simplest operations on earth. For this, however, we should need a script something like Chinese with its ideograms – characters representing ideas rather than sounds. Mathematical notation approaches this ideal and the translation of a mathematical text can therefore sometimes be near to a $C' = C''$ process; yet even mathematical writings use a certain amount of ordinary language, and in so far as they do this, $C' = C''$ is not sufficient.

In a way, this approach is the obverse of the word-name and list-of-words one. According to the first theory of language, every word is a name of a thing in the external world, and the name is the clue to the thing. According to the second, everything in the world does have a name, but the thing is the clue to the name. The former believes that the knowledge of the word as such gives us the knowledge of the thing. Hence translation can be performed entirely in terms of words. But if we look at it from the other side, we need only knowledge of the things, and the words can receive little attention. In either case the word is a label, but the label may be all-important or it may be considered unimportant, a mere convenience. In the one instance all you need is a dictionary, and possibly a grammar book, and in the other you will get by with a small glossary as long as you are an expert on the subject matter.

There is obviously some truth in both these theories, but the greatest truth surely lies somewhere in between. It cannot be denied that translation is a transaction involving words. But so is talking to friends, so is delivering a lecture on the quantum theory or a speech in a play by Shakespeare. To say that translation is a replacement of words in one language by those in another is no more than the truth, and yet it is as informative as saying that speech is the uttering of one word after another. The words are a means of expressing something, of communicating something, and the purpose of the communication is the overriding concern. But words differ from tools in the ordinary sense because they are not the deliberate and conscious creation of man. They are as much a historical development as his social institutions. They are not means at the complete disposal of the individual, completely at his beck and call, to shape as he wishes. They function under strong restrictions and offer some resistance to individual manipulation. Words are means, but means with peculiar properties.

Translation is the transfer of meanings. Words are not necessarily the names of things or of ideas. They have some relation to these, but not a direct or representational one. In addition, they combine with one another, change their forms and follow one another in accordance with rules that vary from language to language. The essence of the matter was stated by the great biblical translator of the fourth century AD, St Jerome: *Non verbum a verbo sed sensum exprimere de sensu* (Not word for word but meaning for meaning).

Thus it can be said that words are a vehicle of communication and express meanings. They cannot be ignored, but they are not all that should be noticed. The rules of grammar govern the manipulation of words and certain principles determine the use of items of vocabulary. The understanding of these – which is often intuitive – is necessary in order to carry out a transfer of the content of a message from one language to another. In addition, extra-linguistic factors are brought into play, together with those that are strictly linguistic. Among these is the subject matter of the text. They also include the objective of the originator of the text – the author – and the intended reader, perhaps the time and place as well.

As a linguistic operation translating is still extremely complex. The transformations of language are not a perfect parallel to those of mathematics. Natural language is a symbol system, as is mathematics, and this is one of the reasons for confusing it with artificial systems like the Morse Code or computer languages.

None of these symbol systems has the complexity and flexibility of natural language. The artificial systems are completely governed by rules, for they are conscious constructions of the human mind, but this is not true of natural language. It is not completely governed by rules, or at least not in any way that is understood at present. It is certainly undeniable that language, some manifestations of language, can be explained in terms of rules. Grammatical rules, for example, can be inferred from observation of the behaviour of language and certain structures can be discerned in vocabulary. But we can do this only to a limited extent, and above all grammatical rules are subject to constant, if relatively slow, change. The rules apply to parts of the language system at any given moment, but not to the whole of it all the time. Modern English has a different structure from the English spoken by King Alfred. There is always a creative and innovative process at work in language and indeed this is evident more today in the language of science and technology than anywhere else. Language is not a finite system. It entails a complex human activity. The artificial languages are finite systems and are derived from the primary natural language.

The two extremes, W' = W" and C' = C", make translation seem a simpler activity than it is. Each minimizes the difficulties. but translation is a meeting-ground for many disciplines. Together with linguistic factors, translation also entails those factors derived from contact between the language system and the outer world, 'from the individual's familiarity with the system, and from the symbolic value which the system as a whole is capable of acquiring and the emotions it can evoke'.[1]

Extra-linguistic factors such as history, culture and ideology are important even when we are dealing with technical subjects. But I must emphasize that linguistics is the keystone. It is linguistics above all that provides the means for understanding the processes that occur in translation and for outlining clear procedures for dealing with problems. The rules of grammar are a good example of this and we can gain considerable insight by analysing some utterances in terms of grammatical hierarchies, interrelationships and vocabulary patterns and seeking possible common ground between languages.

It will be noticed that I have used German and English as the main languages of exemplification. From now on German will be treated as the source language and English as the target language, though examples from other source languages will also be used on occasion.

CHAPTER 4

Units of translation

🆁🆁🆁🆁🆁🆁

TRANSLATION may be defined as a process of finding a TL equivalent for an SL utterance. As noted in the last chapter, this cannot be merely an equivalence of words – a matter which will duly become clearer – and to speak of conceptual equivalence alone would be to ignore important stages and parts of the process. In seeking to determine the elusive notion of 'equivalence' in translation, the guiding principle that suggests itself is to find the smallest identifiable unit that can be matched in two languages.

There are many ways of setting about this. One is to compare translated texts with the originals and attempt to retrace the path the translator has followed. In the normal course of events the translator himself will not be aware of what he is doing. He works intuitively until he encounters difficulties. It is then that we become aware of the value of an analysis of the process that brings to light the nature of the problems and with them the possible solutions. The analytic method followed here is the reverse of the recommended practice for translating. Normally the translator should always study the text as a whole before he begins to translate it. After obtaining a picture of the whole he can break it up into its parts – the size and type of the units will depend on the nature of the text, its length, its difficulty, and not least on the temperament and ability of the translator himself. Instead, the analysis will move in the opposite direction from the fragments to the whole, from the simpler to the more complex. This is, incidentally, analogous to the fascinating history of trial and error which marked research into MT. But it also accords with the principle stated above – that the smallest unit of equivalence is to be determined – and this means testing the smallest units available and working upwards until we arrive at our 'atom' or 'molecule' of translation, if there is such a thing.

The smallest unit of grammatical analysis is a *morpheme*, which is sometimes identical with the word, as in *der* and in *Mann* or in *the* and in *man*. But morphemes may be items that cannot exist independently as words do but only as part of a word, as with *-ung* in *Herstellung* or *-ing* in *talking*.

Consider the expression *die Geräuschempfindlichkeit*. This is taken from a German Bureau of Standards (Deutscher Normenaus-schuss) specification (DIN 4109) on noise control in buildings. Only the noun in the phrase will be examined. It can be broken down into the following elements:

Ge + räusch + emp + find + lich + keit

In practice, the translator may seize on the word as a whole. But if he has never seen the word before, and is required to translate it without any help in terms of what he already knows, he will need to proceed on the lines given below. It supplies the pattern for a procedure used when one is confronted by a difficult expression one has never seen before. Of course, to have arrived at the elements shown above it is necessary to have some familiarity with the structure of German words. But even the most optimistic would-be translator of the easiest text on C' = C″ lines must have some elementary knowledge of SL forms.

Looking more closely at the morphemes in the expression, we see the following:

Ge-

A prefix that converts verbs into the past participle form, e.g. *sehen* (to see) into *gesehen* (seen). This will not apply here, since the element which follows, **räusch*, does not have the form of a verb (it doesn't terminate in *-en*). The prefix also occurs with nouns, verbs and adjectives. It converts nouns into collective nouns. It also converts verbs into nouns that signify continued action. The use with adjectives is less common. It is evident that the prefix has meaning; indeed, it can carry any one of several meanings depending on further information (i.e. the nature of the element that follows it and that it modifies).

räusch

This element derives either from a noun with a range of meanings similar to intoxication, frenzy and so on, or from a verb, *rauschen*, which has among its meanings those of 'rustle' or 'roar'. The second is more likely in the present circumstances and if we take the two elements together as one, we will have the noun, *Geräusch*, of the type derived from verbs and signifying continued action. This gives us a noun meaning 'noise' in English.

-emp-

A prefix that suggests 'opposed, against, in the face of, towards, contrary to'. Since it is a prefix which by definition modifies a following element, we require further information before we can decide on its function.

-find-

A stem occurring in the verb *finden* ('find, meet with . . .'). Again, we can tack the two elements together and derive *-empfind-*, which is the stem, for example, of the verb *empfinden* ('to feel, perceive, experience . . .'). Note that I have taken a short cut here in order not to protract matters too much. Strictly speaking, the meaning of *-empfind-* should be derived inductively by testing combinations of meanings of *-emp-* and *-find-*, but there are inherent limits to what we can do in this manner. The description of these limits is one of the themes of this book.

We now have the combination *Geräusch* ('noise') + *-empfind-* (which has something to do with perceiving, feeling, experiencing).

-lich-

This is an adjective or adverbial suffix designating an essential property. It also converts verbal derivations and nouns into adjectives. Since we are dealing with suffixes, modifiers that follow the element modified, we have the necessary information to judge which modification is necessary. In this case we have a verbal stem, *-empfind-*, and can assume that a conversion into an adjective has taken place.

We now have the combination *geräuschempfindlich* (the property or quality of perceiving, feeling, experiencing noise).

-keit

A suffix that converts adjectives into nouns. It should indicate the state of having a certain quality. We have now arrived at the full expression: *Geräuschempfindlichkeit*, which should yield literally: the state of having the quality of *noise in the face of (or: against) find. Or if you like: *noise + in the face of + find(ing) + ly + ness. This is nonsensical. Yet by adopting a method of building up identifiable words where possible, we have arrived at a basic combination: *Geräusch* + *Empfindlichkeit* (noise + the state of having the quality of perceiving). The relation of this combination to an expression like 'noise' + 'perceptiveness' should be clear without our straining any effect. Once we achieve this, it should not be difficult to arrive at the better English form 'noise sensitiveness'.

In the second word, we recognize that the morpheme *-keit* finds a ready English counterpart in *-ness*, which also serves to convert adjectives into nouns. For *-lich*, the English counterpart is the suffix *-ive*, which also converts other forms into adjectives.

It will be noticed that there is a certain hierarchy in the forms we have examined. The word is always on a higher rung of the ladder than the morpheme; where the morpheme has a range of possible functions and meanings, the word is the main element that decides which of these is to be selected. In the German noun, the two main components, which are themselves nouns, are determined by the total compound. Similarly, the components of the phrase (*die Geräuschempfindlichkeit*), article + noun, are determined by the total phrase. The higher rank always determines the elements. The sentence determines the meaning of the phrase. The whole text determines the meaning of the sentence. The situation in which the text occurs determines the meaning of the text.

By this method of passing from the smaller to the higher unit we would arrive in this instance at nonsense, or at best at a very clumsy expression – something like *noise experiencingliness*. Taking the compound as two words, we would have the result suggested, *noise sensitiveness*, which is not apt since 'sensitiveness' is usually attributed to human beings and indicates an emotional state. But *sensitivity* can be used not only for human beings but also for inanimate objects, and *noise sensitivity* is therefore the better translation. The preferred equivalent, however, is that accepted by the experts in the field: *noisiness*. This is quite a big leap and suggests a higher level of transfer than morphemes. In fact the only equivalence at morphemic level is that between *-keit* and *-ness*. These two will generally match up but this is not true of *-lich* and *ive* since the latter has a much more restricted use than the former.

The rule for translation that can be inferred here is that the smallest unit of equivalence is sometimes the morpheme, but this is not always so. In more general terms, the morpheme cannot be used as a basic unit of equivalence, though it is a clue to meaning.

Another example will take the matter a step further. The phrase *die Anschlussleitung* is taken from another specification document. The equivalent preferred by the subject experts in this case was *service pipe lines*. Comparing the SL form with the TL, we notice:

SL	TL
1 article (definite) indicating: feminine gender, singular, nominative or accusative case	1 absence of article
2 noun abstract, singular, composed of: adjunct (N *Anschluss*) + main noun (N *Leitung*) singular	2 noun phrase comprising: adjunct (N functioning as adjective: *service*) + adjunct (N *pipe*) + 3 head noun (N *lines*) plural

The previous example displayed a condensation of the structure of the head word in the phrase *die Geräuschempfindlichkeit* (analysable into two words), whereas the second shows an expansion. Note too that the English counterpart of the first phrase is a single word, but the English counterpart of the second phrase (two words) is three words. In addition, the SL expression provides information that is absent in the TL (gender, case). It is constructed differently in terms of grammar and also as an item of vocabulary. The unit of equivalence here is undoubtedly not the morpheme, nor is it the word. The elements of *Anschlussleitung* immediately demonstrate this: *An-* (preposition with a wide range of meanings); *Schluss* (noun with a wide range of meanings, including: 'closing, conclusion . . .', related to verb: *schliessen*, which with preposition *an* in a phrase like *schliessen an* can have the sense of 'join, connect'): *Anschluss* (connection); *leit-* (stem of verb *leiten*: 'lead, conduct . . .') + *-ung* (suffix that converts verbal actions into nouns), hence

Leitung ('conduction'). This would yield something like 'connection' + 'conduction', and again there is a leap to the preferred equivalent. In other words, equivalence here is neither morphemic nor on the word or phrase level, but involves some unit higher up in the hierarchy.

The need for a higher unit of equivalence can be shown even with a much simpler phrase, one as simple as *der Mann*. True, this can be replaced in English by *the man*. It has all the appearance of an ideal case – a one-to-one correspondence of all the minimum elements, in this case words (morphemes) with identical syntactic functions: article (definite) + noun (animate: human being). But suppose we now have to find an equivalent for the phrase *den Mann*. The result is still *the man*. But the structure of the German phrase has changed. The nature of the article is different; it indicates something different. Where *der* marked masculine gender, singular and nominative case, *den* marks masculine gender, singular and accusative case. The first article tells us that the phrase must be the subject of a sentence, the second that it must be the object. The English phrase cannot indicate gender and does not carry any information as to whether it is the subject or object of the sentence. We need an entire sentence to know this. The unit of equivalence is therefore not the word but the sentence. Only at this level will we know the English word order and therefore the function of the phrase.

To summarize the findings: the smallest unit of equivalence can vary considerably and is rarely the morpheme. A word in one language may be the equivalent of a phrase in the other. These two points are of very great importance. Our examination so far has of course been in terms of grammatical rank and function. It has also been in static terms. The equivalent unit has been viewed rather as a mirror image of the original unit.

The weakness lies not only in the size of the unit but in the static approach as such. The text is seen as a document that might just as well be written in a dead language. It is a product, an artefact, which is complete and finished. But in fact translation, as we have seen, is a part of the communication process, and communication processes are not static. It is a procedure in which one person transmits information to others. There is intention on the part of the sender to achieve some effect in relation to his intended reader. The equivalence that matters, then, is one of effect. This can be described in terms of the sender achieving the same effect on the reader with the TL text as would be achieved if the reader were able to read the SL text. In technical translation, the emphasis will be placed on the effect on the reader.

Within this framework, the relation of text to text (SL text to TL text) should not be one of replication but a dynamic one in which the existence of two texts should offer no obstacle to communication between sender and reader. The translation should be transparent in the sense that the intention of the original is clearly evident through it. The use of language – the establishment of linguistic parallels – is subordinate to the creation of the equivalent effects. In this sense, the appearance of *service pipe lines* as a translation equivalent for *die Anschlussleitung* is justifiable, although from a static, replication viewpoint it cannot be justified. The static units of equivalence are useful for understanding the processes and may be a convenient instrument for achieving the desired effects when used correctly. But the unit will vary. The smallest possible linguistic item, such as the morpheme, may be the static equivalent in one context. In another context the sentence may be the static equivalent. In yet another, it will be the entire text. It is often necessary to go outside the text to the subject field as a whole. Subject knowledge can on occasion make short cuts, a C' to C'' transition, possible; usually dynamic equivalence, it is also most likely in the sphere of terminologies. The classic cases would be mathematical and chemistry texts. The conceptual leap does not help with the other, general linguistic elements that contribute to the make-up even of technical sentences.

The importance of the principle that a dynamic account of translating is preferable to a static one and that, in any event, the larger the unit of translation, the more accurate the equivalence, is well illustrated with reference to three main pitfalls. These are ambiguity, interference and variation in range. A glance at these – fuller treatment follows later – is highly instructive.

Ambiguity, for instance, can result from vagueness of expression or from the use of words or phrases with several meanings. multiple meaning, *polysemy*, is a major issue in translation. If a foolproof rule could be devised for solving this problem, for selecting the single correct meaning in a given instance, we should be well on the way to establishing a science of translation, with momentous results for data processing by means of the electronic computer and for linguistics in general. The special form of the multiple meaning problem in translation is known as *homography* – one form of the written word has several possible meanings – as distinct from *homonymy* – one form of the spoken word has several possible meanings. We find homography in words like *Schloss*, an example we have already noticed; *lead* (noun or verb, metal, dog's lead, clue); *mean* (adjective, noun, verb; *stay* (noun or verb, *stays, stays of execution, a long stay*). Examples of homonymy occur in expressions like the following: *a red book* and *a read book*. On the syntactic level, a sentence that is by now quite famous in linguistic circles is: *Flying planes can be dangerous* (to fly aeroplanes can be a dangerous pursuit; aeroplanes [or planes used in carpentry?] that are travelling through the air can be dangerous).[1]

There can be no doubt that in the majority of these examples the minimum unit of equivalence is the sentence. The ambiguity of *lead* may be resolved on the phrase level with *lead pencil* or *lead weight*. But *a red book* may be distinguished from *a read book* only on the sentence level, while the flying planes example needs a larger unit still.

Another source of confusion is *synonymy* – several words with the same meaning – as with *doctor, physician, medical man*. Medical terminology in particular is rife with synonymy, but it occurs in many other technical contexts. In general, technical language seeks to avoid synonyms or homographs and to achieve the ideal of one word/one meaning, but it would be idle to pretend that this has in fact been accomplished, and that there are not many technical synonyms, homographs and homonyms. For instance, German has both *Atom* and *Kern*, just as we have *atomic* and *nuclear* in English. Both *Atomenergie* and *Kernenergie* can be translated as *nuclear energy* (which has become more or less established as the correct technical term).

But this is a minor matter compared to the troubles that polysemy can produce. A very interesting example of this is the German word *Leistung*, which has a high frequency in technical texts. *Leistung* can mean: efficiency, power, capacity, output, life, accomplishment, achievement, result, obligation.

Antriebsleistung	propulsive force (milling engineering)
Mahlleistung	grinding efficiency
Leistungsgrad	performance

Leistungen	capacity
Leistung	$\begin{cases} \text{power (steam engine)} \\ \text{capacity (machine)} \end{cases}$
Spitzenleistung	(output) peak output (engine)
Spanleistung	cutting power
Schnittleistung	cutting power
Hochleistungsmaschine	high-duty machine
die erzielten Leistungen	the results obtained
Leistung	$\begin{cases} \text{efficiency (human beings)} \\ \text{achievement (human beings)} \end{cases}$
elektrische Leistung	electric power, power
abgegebene Leistung	output (electricity)
aufgenommene Leistung	input (electricity)
Leistungen	obligations (commerce)

The relevant meaning of *Leistung*, as can be seen from these examples, can be distinguished to some extent by means of the specifying element in the word (*ANTRIEBSleistung*, *MAHLleistung*, *SPANleistung*), or a specifying adjective, as with *elektrische Leistung*. But it should also be clear from the examples that this is often insufficient, and that the translator must be familiar with the subject field. This knowledge must moreover frequently be very specific, as with the generic expression *Leistung*, which is a very common term in engineering contexts. Hence the particular usage must be carefully distinguished in relation to 'steam engine' or 'human being' or 'electricity', all of which come within the field of engineering.

The sentence is probably the typical unit of translation equivalence, but only in the sense that it represents the most convenient collection of items to work with, providing we already know the more general context, the subject field. In practice, the translator will be aware of what the translation is dealing with in general (engineering, toxicology and so on). He may also know the more specialized aspect of the field which the translation is concerned with (e.g. a specific disease found in vines in a district of Italy). Within this framework he will perhaps translate sentence by sentence, and proceed by breaking up the whole into smaller units. But the human translator has the invaluable ability to think not in terms of replicating units of text but of creating analogies that will convey the intention of the original author. In general, the translator tends to replicate when he has the source of the utterance in mind, but to analogize whenever he is more concerned with the reader.

Where knowledge of the subject matter is necessary, the equivalent unit will lie on the conceptual plane. The principle adopted as between seeking replication or analogy applies here. The lowest hierarchical level does not yield the answer found on the higher level. Some ambiguities are resolved on the grammatical level, others are not. Where the ambiguity can be resolved on, say, the conceptual plane, an approach purely in terms of grammar will be a dead end. But an approach on the conceptual level can also bring into account context and grammar.

The phenomenon of interference is the persistence in the TL of usages peculiar to the SL. If there are corresponding usages in the TL, then interference will not be observed, but such correspondences exist only occasionally, depending on the pair of languages in question. A type of interference that is notorious in translation practice – it is of course prompted by a desire for $W' = W''$ transfers – is the one called in French *faux amis*, 'false friends'. Such expressions have a similar appearance in both languages but in fact have different meanings. A transfer of SL word order into the TL may be involved, but the confusion caused by false similarities in forms of words and phrases is even more common. It is especially frequent where there has been a great deal of borrowing between two languages. The usual course of events is that the borrowing in language B develops different meanings from the original in language A. This is found very frequently between French and English. A typical example is the translation of *éventuellement* (possibly) as *eventually*. German has a similar use of *eventuell* to French; so have Swedish and a number of other languages, and the unwary translator is very likely to confuse it with *eventually*, which does not convey any notion of contingency but has rather the sense of 'ultimately'.

A blatant example (perhaps due to fatigue, for it is an error made by a capable translator whose mother tongue is English) is:

F *afin d'éviter les déperditions calorifiques*
E **in order to avoid calorific perdition* (instead of: *in order to avoid heat loss*)

Here is another example taken from a German technical magazine:

D *Natürliche und zivilisatorische Strahlenbelastung des Menschen*
E **Natural and civilising radiation exposure for humans.*[2]

A more satisfactory rendering would have been: *Human exposure to radiation from natural and man-made sources.*

The cause of these aberrations may be fatigue, inadequate linguistic knowledge, lack of subject knowledge or a combination of these and other deficiencies, but what is interesting about examples like these is that they are all the result of an attempt to replicate the SL items, at the level of words or even smaller elements. This is in contrast to the 'noisiness' and 'service pipe lines' examples, where the transfer occurred on the conceptual plane.

The problem of units of translation is a difficult one. It is particularly so for MT, but also in human translation is so far as the design of aids for the translator are concerned. In other words, it relates to the question of what should be put into a dictionary intended for translators rather than for the general public. This is an important element in the discussion of procedures used in translation and techniques for facilitating the task. In fact between any pair of languages that have been in contact for some time we have what might be called 'translation paths'. Certain words, certain phrases and indeed certain concepts are translated so often that they become established. Translation clichés arise. From experience, the translator knows that a certain expression is best translated in a certain way. There are numerous fixed equivalents of this kind in technical languages. Here the translation unit is smaller than the sentence; it is usually the word or phrase.

The procedure should, however, be to seek the conceptual equivalent, to define it more accurately, and then to render it in the linguistic terms of the TL. In the hierarchy of translation equivalents, the top rank is that of the concept, followed by the lexical and syntactic equivalents on the level of the sentence, and then the smaller units – phrase, word, morpheme.

TYPES OF EQUIVALENCE

The different types of equivalence can be tabulated in the following way:[3]

$$G' = \text{source language structure}$$
$$G'' = \text{target language structure}$$

$$M' = \text{SL meaning}$$
$$M'' = \text{TL meaning}$$

1   *Structural similarity + meaning similarity*
$$G' = G'' \qquad\qquad M' = M''$$

This is an ideal case, most likely to occur in languages that both are genetically close and have experienced considerable communication between the members of the two speech communities, as in some bilingual countries. It is also more likely to occur at the lower levels of the hierarchy, and less likely to occur at higher levels. The more complex a sentence, the less likelihood there is of such complete convergence.

Thus *er ist* becomes *he is*, a case of $G'M' = G''M''$. But with a slightly more complicated construction we have divergence in structure: *Er ist gelaufen*: *he ran*.

2   *Structural similarity + dissimilarity in meaning*
$$G' = G'' \qquad\qquad M' \neq M''$$

Examples of *faux amis* generally come into this category, one being the 'radiation exposure' sentence cited above. A literal translation may be of type 1, but there is also the risk of it being of type 2.

3   *Structural dissimilarity + similarity in meaning*
$$G' \neq G'' \qquad\qquad M' = M''$$

An example here would be the corrected version of the 'radiation exposure' sentence. Translations on the conceptual plane come into this category, even if the transfer is at a lower level in the hierarchy than the sentence (note *die Geräuschempfindlichkeit* and *die Anschlussleitung*).

4   *Structural dissimilarity + dissimilarity in meaning*
$$G' \neq G'' \qquad\qquad M' \neq M''$$

This category includes completely incorrect translations, and utterances that are impossible to translate.

The ideal in translation is to achieve an equivalence of the first type ($G'M'' = G'M''$), but this is rare. The next best, and generally the best in practical conditions, is of the third type ($G' \neq G''$, $M' = M''$). A 'rough' translation can vary from the worst example of the fourth type (completely different in all respects from the original and so utterly unsuccessful), to the third type, perhaps with a clumsy form of expression, but conveying the meaning.

SCALE OF EQUIVALENCE

We can also make a scale of levels of equivalence in ascending order from the substitution of the simplest linguistic signs to more elaborate groupings. These levels are all encountered in practice.

1 *Substitution of printed letter for printed letter*

Transliteration is generally required for titles, quotations and references, and usually by research workers, librarians and documentalists. It is necessary when the SL uses a different script from the TL, and with Russian (Cyrillic alphabet), Hindi (Devanagari alphabet), Arabic, Hebrew, Japanese (katakana and hiragana). A familiar example of transliteration is the word *sputnik*, taken from Russian. There are numerous systems of transliterating and little consistency is shown. Virtually every language has its own system for rendering Cyrillic script and Japanese and Chinese ideograms. German may transliterate Cyrillic з by the letter *S*, whereas most other languages use *Z*. The Cyrillic ш is rendered in English by *sh*, in French by *ch*, in German by *sch* and in Dutch by *sj*. In 1955 the International Standards Organization made a recommendation (ISO/R9) which has been reprinted in British Standard 2979:1958 for a system called 'Latinica' (pronounced Lahteenitsa). The symbol for ш in Latinica is š. The aim in transliteration is to reproduce letter by letter, not to reproduce pronunciation. This is rather difficult in the case of Chinese, which has no alphabet and which is a language in which tones are meaningful. Efforts are now being made to simplify the writing of Chinese, with the use of a Romanized script, for instance, and the same applies to Japanese.

Even languages with the same type of alphabet may display variations, since some have letters not possessed by others, and using diacritical signs. A typical diacritical sign in Spanish is the tilde, the ∼ sign placed above the letter *ñ*, which gives it the value of *ny* as in *canyon* (which in Spanish can be written *cañon*) or *mañana*. German uses the umlaut over vowels, as in *Mädchen, grösser, über*; these vowels are sometimes reproduced in English as *ae, oe, ue* (*Maedchen, groesser, ueber*).

Close to this level are equivalences that are easily recognizable as being phonemic, such as Spanish *vatio* for English *watt*. We may take this a bit further and note that German *V* has the same sound as English *F*. This is observable in the correspondence *Vater*: *father* (which incidentally also draws attention to the lack of the *-th* sound in German). *C* occurs in German only in borrowings from other languages, and *K* may be used where English would have *C*, as in *Kaffee*: *coffee*. The letter *J* has the value of consonantal *Y* in English, as in *Jahr*: *year*.

2 *Substitution of morpheme for morpheme*

This occurs fairly frequently in technical terms where the main element is an internationalism. In general, only certain morphemes lend themselves to this level of equivalence between German and English. One is *-er*, as in *Arbeiter*: *worker*.

3 *Substitution of word for word*

4 *Substitution of phrase for phrase*

Where a stock of equivalent phrases has been built up for two languages, it is possible to achieve a transfer of the $C' = C''$ type. The stock translation equivalents have become labels – we could even give them numbers – and the transfer bypasses the word values, proceeding from idea to idea. This of course occurs only in hypothetical circumstances, since it is not realizable to any great extent in practice, because of the way language is constantly changing. But to some extent a translator can build up a stock of phrase equivalents, a sort of private dictionary.

5 *Substitution of sentence for sentence*

This seems to be the most effective unit of translation but, as we have already seen, there are occasions when smaller units are possible, and others where the sentence will be too small a unit.

6 *Substitution of a context larger than a sentence for a similar one*

7 *Substitution on the rank of situation, i.e. having recourse to extra-linguistic data*

It should be noted that substitutions between units of different size may be necessary: a word in the SL may need a phrase in the SL, for example, and vice versa. What can be said with one word in one language may require a great deal of circumlocution in another, and in certain cases we may only be able to draw the reader's attention to the situation in which he will be able to experience and understand what the author is aiming at. We say in circumstances like these that the expression is untranslatable. This happens a good deal in literary translation and is not unknown in technical translation.

CHAPTER 5

The language barrier

🔯🔯🔯🔯🔯🔯

THE language barrier results from the fact that each language is
an individual and distinctive system, that it has a structure and is
not simply a list of names. There are other barriers to understanding a communication in a foreign language, but the reader is presented in the first place with a language barrier, owing to the
difference in structure between one language and another. Not all
structural features are understood equally well. Usually structure
is understood to mean grammatical structure, in which a degree
of regularity and pattern in the organization of the language can
be discerned. Vocabulary also has a structure, but this is a more
complex matter and one that is imperfectly understood. The structure that is evident in both grammar and vocabulary may in some
respects be shared by more than one language, but there is never
a complete parallelism between any two languages, no matter how
closely related they are. Indeed its distinctiveness and the difference between it and other languages are essential in defining a
language as such.

Examples of differences between languages are that language A
will have grammatical categories that are absent in language B;
language B in turn will have grammatical categories that are
absent in language A; language A will have vocabulary that is
absent in language B and vice versa; one language will use its
grammatical categories and vocabulary in a different way, in
different combinations and patterns, from another. There will also
be disparities on the grammatical level and on the lexical level,
and, most probably, one language will have no words for a concept expressed in the second language.

Quantitatively, there will be disparities on each of these levels,
as follows:

$$A\ (items) \qquad B\ (items)$$
$$1 \qquad\qquad \emptyset$$

A (*items*)	B (*items*)
I	2 or more
Ø	I
2 or more	I
Ø	2 or more
2 or more	Ø

One-to-one correspondence on all levels is a rare occurrence. It is the ideal condition for word-for-word translation. Where it occurs on only one of the levels, some sort of language error or clumsiness will result: foreign-sounding turns of phrase or pidgin of some kind, or ideas that are alien to the TL speech community. A one-to-many correspondence will at best require only a grammatical adjustment in the TL, but a one-to-none correspondence will certainly require the process I have called analogization. The problem for the translator is entirely that of the one-to-none and one-to-many correspondences. For instance, there is no counterpart in Europe to the American *drugstore*. Such a shop does not exist outside North America and there is no word for it. In English-speaking countries that have no such shops the use of the word is highly misleading, unless its specific American context is implied.

A word may exist in another language, but it will rarely have the significance it had in the original one. Spanish *hidalgo* is an example. Another is English *gentleman*: the word is known in other languages, but it is not used in the same way and with the same meaning as in English, any more than *drugstore* is outside North America. Yet most English-speaking people probably understand the idea of the American drugstore perfectly well. Words expressing abstract ideas present an even bigger problem and great care has to be exercised in discriminating between the distinctively French meanings of, say, *culture* and *civilisation*, German *Kultur* and *Zivilisation* and their English counterparts. Accurate translation of these expressions inevitably requires circumlocution – one-to-many correspondence.

A comparison between a pair of language systems, such as German and English, results in a general picture of their similarities on a number of levels. This comparison is implicit in the mind of the translator. To make this process possible the two languages must have something in common. There must be overlapping of categories, items, functions and so on. This is true of all languages. If it were not, a science of language could scarcely be possible, and certainly no translation could in principle be possible between any two languages. These disciplines are based on the existence of universals in language.

The first and most important universal, or common feature, is that all languages have a structure (grammatical rules) and follow certain patterns of vocabulary. If these patterns were fully understood especially in vocabulary, an exact science of translation could be developed. Nevertheless when we attempt to translate we assume that the SL is made up of sentences in which we can recognize something like a subject and a predicate. This process is relatively easy between the main Western European languages. Both German and English, for instance, have word classes; they form a number of similar sentence types; and there are resemblances between a large number of lexical items. The definite article in German overlaps in its function with the English, but it has a wider range of application. This includes the ability, one that is not possible in English, to specify differences in meaning, as in *der See* (masculine): *the lake* and *die See* (feminine): *the sea*. Underlying the apparent similarities between the German and the English definite article is a general many-to-one correspondence.

The range of application of any particular feature in the pair of languages varies on a number of different levels, which for convenience are arranged as below:

1 grammatical	5 denotative
2 morphological	6 connotative
3 syntactical	7 cultural
4 lexical	

The first four are linguistic levels, while the others relate to the external world. The first three levels are grammatical but are not arranged in order of ascendancy; nor are the last three levels, which may be called levels of situation, arranged in a hierarchical order. But the broad levels – grammatical, lexical and situational – do form an ascending order. The morphological and syntactical ranges are grammatical, but the term grammatical is used here in a narrower sense to include all grammatical phenomena that would not be classed as morphological and syntactical. The lexical level overlaps with the denotative and connotative levels and the last three also overlap. The differences, however, are very clear, as will be shown.

GRAMMATICAL RANGE

Grammar in the broad sense is the inventory of forms in a language and the rules according to which they can be combined. We are concerned here chiefly with the inventory. The expression 'grammatical' in the restricted sense covers categories such as tense, number and gender; word classes such as noun, verb and so on; structures such as co-ordination and modification; and rank of expressions (word, phrase, clause, sentence).

There may be a lack of correspondence between categories (one-to-nought and one-to-two) and functions of categories. Points of difficulty between German and English include the functional range of certain word classes, as with the article exemplified above, differences in adjectival comparison, verb categories, and certain other features.

The inflectional system of the German verb offers a most striking contrast with the English verb. English does have inflections, but these are few. An example like *nehmen*: *to take* shows the difference.

present tense

ich nehme	I take
du nimmst	you take
er/sie/es nimmt	he/she/it takes
wir nehmen	we take
ihr nehmt	you take
sie nehmen	they take

English has only two endings: the base form that can be described as *take* (German counterpart *nehm-*), with the endings ø as in *take* + ø; and the form *take* + *s*. The German verb, however, has four endings (*-e*, *-st*, *-t* and *-en*). There is also the internal modification in the second person singular (*nimmst*) and third person singular (*nimmt*). As a case of many-to-one correspondence, it is less likely to offer difficulty to the translator than its reverse. We have more information than is needed in English.

A more serious problem is the lack of parallelism between German and English tenses. In particular, the English progressive forms have no counterpart in German and translators often experience difficulty in rendering the German expression exactly in English. This is a none-to-one and one-to-many problem.

Other disparities are the overlapping between adjectival and adverbial functions in German, and the more frequent use of prepositions in English to indicate relationships that in German would be marked by a case relationship (e.g. *dem Mann*: *to the man*). Adverbs are used to qualify verbal forms in German so as to distinguish between definite tenses and progressive tenses. *Ich gehe* can be translated either as 'I am going' or as 'I go', a one-to-many correspondence. But *ich gehe schon* can be replaced only by *I am going* and *ich gehe jeden Tag* by *I go every day*.

The modal forms produce a great deal of difficulty: for example, the uses of *shall* and *will*. The range of *sollen* partly overlaps with that of *ought to*, but phonemic interference (identification of similar-sounding words) frequently occurs, with the result that it is confused with *shall*. This is encouraged by the overlapping of *ought to* and *should*. There is also confusion between *müssen* (compulsion) and *dürfen* (obligation). The *may* in 'You may not walk on the grass' corresponds to *dürfen*, while in 'I may walk on the grass tomorrow' it is the equivalent of the auxiliary *kann* (possibility).

MORPHOLOGICAL RANGE

This mainly concerns similarities and differences in the formation of compounds and derivational series. An example of the former is German *Atombombe* and English *atom bomb*. Where German usually combines the two nouns to form one, English associates them more loosely as separate words (noun acting as adjunct and noun acting as head-word). Where the combination is a closer one, as in *blackbird*, the sense of the original modifier is often blurred, which is not the case in German. Thus *blackbird*, like *cranberry*, is felt to be a single integral word and not two words put together.

Similarities in derivational processes are exemplified in a form like German *lieblich*, which has a possible English equivalent in *lovely*; but -*lich* does not always correspond to -*ly*, as the example of *kleinlich*: *petty* shows. The expression *die Geräuschempfindlichkeit* (*noise sensitivity*) is another illustration of the difference in morphological range, since German -*lich* is replaced by English -*iv(e)* and -*keit* by -*ity*. The variation in range between German -*keit* and

English forms can also be seen in, say, *Ordentlichkeit*, which can be translated as *orderliness* or *regularity*.

Similarly, prefixes do not always match: *wiederaufbauen*: *to re-build*; *wiederbringen*: *to restore* (*wiederherstellen* can also be translated as *restore*); but *wiederkäuen*: *to ruminate*.

SYNTACTICAL RANGE

The contrasts here are between sentence types, the construction of clauses and sentences and the order of words within sentences. Word order is the most noticeable feature. A subject-verb-object order is common in German and will normally find an exact counterpart in English.

Bild 1 zeigt das Schema eines Linearhubmotors....
Figure 1 is a diagram of a linear lift motor....
(literally: Figure 1 shows the diagram of a linear lift motor...

A different word order is possible in German, but not in English. The above sentence follows the S-V-O pattern, but German can also have O-V-S patterns. In an English statement sentence it is obligatory for the subject to precede the main verb form and for the object to follow it.

LEXICAL RANGE

From the previous remarks on value and designation, it should be clear that apparent correspondences in vocabulary between different languages are only partial. We have noted that *Fleisch* can be replaced in English by either *meat* or *flesh*. Every expression in a language has a number of possible meanings, only one of which may come into play in a particular context. The range of possible meanings may be very great, or it may be limited. In technical texts, as we have seen, a word like German *Leistung* has so wide a range of possible meanings that it has to be made more specific by the context, or by a modifying expression such as an adjective or another noun. When it refers to the efficiency of a machine in terms of the degree of energy transformation, it is usually rendered in English as *efficiency*; when it means not the amount of work but the output capacity, the capacity or the output – the work done in terms of a time unit – it is translated as *power*; when it is characteristic of the yield of a machine, it may correspond to *capacity*, *production capacity*, *productive capacity*, *output*; when used of oils, it is translated as *serviceableness*; in regard to a human being it becomes *efficiency*, *achievement*, *accomplishment*.

One language may make distinctions that are unknown to others. For instance, German has the expression *die Blume* (the flower) but also *die Blüte* (the flower, blossom, bloom) whereas the Romance languages have only one name (F *la fleur*, I *il fiore*, S *la flor*). Whereas *Leistung* has a number of designations (references to the real world) that can be selected according to the context, the designation for the object 'flower' is spread over two expressions in German.

It is not only the lexical range of an expression that is important. Its collocational or co-occurrence range is equally important. Certain words cannot be used in association with other words. As we have seen, restrictions of occurrence prevent *meat* from being associated with *wound* in English, and make it obligatory to speak instead of a *flesh wound*. *Flesh* collocates with *wound*. One of the definitions of *wound* is that it is normally used only of living beings. Similarly, *injury* is normally used only of living beings. Only by using the expressions metaphorically can we extend the range of association, as when we talk of a cliffside being wounded by mining operations and displaying huge scars. By using *wound* in this manner we are defining the inanimate cliff as a living creature. This is a common occurrence in primitive cultures and many traces of pre-scientific, mythical beliefs remain in our modern languages.

Restrictions of collocation are particularly pronounced in greetings. We can say in English *good day* (or *good morning* or *good evening*), but we never say *excellent day in this way, or *bad day, *reasonable day or *fair day. There is no logical reason for this, it is simply a custom that *good* goes with *day* in a greeting, while other adjectives do not.

These restrictions seem to belong purely to the realm of vocabulary, as distinct from the external world. *Good day* is a purely lexical association; otherwise *excellent*, *fine* or other similar adjectives could be substituted. Yet we speak of *white paper* only when a sheet of paper is white in colour. This is a distinction that belongs to the external world. We can call it *white paper*, *das weisse Papier* or *le papier blanc*; it is still paper that is white in colour. Of course most paper that we use for writing is white and therefore there is a high probability that the colour term we use for paper will be 'white'. But if we leave it open as to whether it is writing paper, wrapping paper or any other sort of paper, it does not have to be white. Then in the expression *White Paper*, used to describe an official document, the association has become a lexical one. It does not mean that the document is necessarily white or even that it is on paper – nowadays it could be on tape or microfilm. In German the corresponding term is a 'book', *das Weissbuch*. White can equally be applied to horses, to snow and to a number of other things. In the case of horses, other colours can be used. This is a matter of the real world. But when *blond* has the restriction that it can be applied only to hair or complexions or lace, we are faced with a restriction on lexical co-occurrence.

These co-occurrence restrictions, especially those of a lexical nature, represent some of the greatest difficulties translators have to face. They vary from language to language, sometimes overlapping, sometimes not, and a considerable mastery of a language is needed to distinguish between free combinations (from a lexical point of view) such as *white paper* and *white snow* and restricted ones (again from a lexical, not a conceptual point of view) such as *blond hair*.

DENOTATIVE RANGE

This refers to the material world. A given language may not only analyse the external world differently from another in having one expression like *Fleisch* for an item where another has two, *meat* and *flesh*, but the objects that appear to be represented by equivalent or near-equivalent expressions may in fact not be identical. A language may have no expression for the item because it does not exist in that language and speech community, or because there has been no need for it. When the need arises the gap may be filled by a term borrowed from another language or by the creation of a new term. A large number of expressions have been borrowed by modern European languages from Greek and Latin and these two languages are still the most common source of new scientific terms. The word *science* is itself of Latin origin (*scientia*: *knowledge*), while *technology* comes from Greek *tekhnologia*. Sometimes the idea alone is borrowed, rather than the actual word. A new word is then coined in the borrowing language to express the concept.

Gewissen was introduced into German on the model of Latin *conscientia* (*conscience*). This procedure is common in German. The result is a host of German technical terms that differ from those in English and the Romance languages, which are direct borrowings from the classical languages. Yet German also has a series of classical terms alongside the local coinages, as with *Geographie* and *Erdkunde* (noun), plus the adjectives *geographisch, erdkundlich*.

The item in question may be a concept, a product, a process, an institution or a custom (the latter are dealt with under the heading of cultural range). It may be material, conceptual or cultural. For example *Föhn* is a wind for which there is no corresponding English condition, so it is simply translated as *föhn*; as we have seen, *drugstore* is found in North America but there is no such thing in Europe.

The distinction to be drawn is a state of affairs in which a) the subject exists on both sides of the language barrier and there is a one-to-one correspondence between the words designating it (e.g. D *Haus*, E *house*); b) there is one-to-one correspondence between objects but different words are used (e.g. D *Röntgenstrahlen*: E *X-rays*); c) there is a two-to-one correspondence between objects, even though the terms may seem to have one-to-one correspondence (e.g. D *Wissenschaft*, E *science* [*Wissenschaft* covers the humanities and social sciences, whereas *science* has come to mean only the physical sciences]); d) there is a one-to-none correspondence, as with certain institutions and concepts (e.g. D *Blitzkrieg*, which was borrowed by English); educational concepts like F *baccalauréat* and D *Abiturient* come into this category.

Denotative range is a more inclusive term than either connotative or cultural range.

CONNOTATIVE RANGE

Certain expressions may be translatable even on one-to-one terms, yet the translation will still be wrong because it does not take account of the historical associations of the expression. For example D *Kultur* can be rendered as E *culture*, but it will almost always create the wrong impression because the word 'culture' does not have the same emotional associations as *Kultur*. This is immediately clear if we use the German word in an English context. It carries, to say the least, an intensity that 'culture' does not possess. There is no one-to-one English equivalent for many of the derivatives of *Kultur* such as *Kulturgut* (literally 'cultural possessions') or *Kulturkreis* (literally 'cultural circle', but applied to an area with important common cultural traits). The word *Volk* is another that has no real English equivalent. The literal rendering is 'people', but this does not have the evocative power of *Volk* in German. 'Nation' is not a true substitute, although on some occasions it will be the most appropriate equivalent.

CULTURAL RANGE

A speech community has historical associations and institutions peculiar to itself, and even within such a community some institutions are restricted to the members of a particular sub-group, such as a religious denomination. *Drugstore* is again an example within the English speech community.

Many countries have presidents. But *the President of the United States* does not correspond in role and function to *le Président de la République* in France, which also has a prime minister, to the *Bundespräsident* in Germany or to the *State President* in South Africa, who is a purely ceremonial figure. In technical translation the problem is largely one of rendering academic and professional titles, although heads of state and legislative bodies do also figure in texts.

Another problem, as will be shown, is that one language may use a concept on a higher level of abstraction than another and there is a gradient between the two closest equivalents. The example of *Wissenschaft* again illustrates this, as it is a more general term than English *science*, so that in most contexts the true equation is D *Naturwissenschaften* = E *science*. *Natural science* is, of course, also correct, but it does not stand in opposition to *science* as *Naturwissenschaft* does to *Wissenschaft*. We often find, too, that where German will use an expression belonging on a more formal level, the more usual equivalent in English is colloquial. In German *Zündkerze* (literally 'ignition candle') will be the most common term, whereas in English *sparking plug* or *spark plug* is used rather than *ignition plug*.

A crucial matter in translation, and one of the most difficult of all, is idiom. With this we can associate metaphor, which is a very frequent phenomenon in technical language. The terminologies of all disciplines are full of metaphors (e.g. 'breeder reactor' in nuclear physics). Idiom is recognized as a fundamental problem in MT. The example of *red herring*[1] has been used as compared with *red tie – er trug eine rote Krawatte*: *he wore a red tie*, but *Finte* (*Ablenkungsmanöver*: *feint, trick*) for *red herring*, not *rote Hering*.

In this chapter we have surveyed the problem of finding matching forms in two languages.[2] The following pages will deal with this problem in greater detail.

CHAPTER 6

Grammatical range

҉҉҉҉҉҉

A translator needs grammatical knowledge to enable him to resolve ambiguities due to structural features and to foresee and avoid interferences between the two languages. It should also enable him to distinguish between optional and obligatory forms in the language, those things that have to be said and those that have alternatives. This kind of knowledge is second nature to all of us as far as our mother tongue is concerned. It may be very deeply rooted in a person with a good command of another language but he will never be able to take it for granted in the same way. In any event, nobody has perfect control of any language. This would mean not only carrying in one's head the entire vocabulary and all the grammatical rules of that language, but being fully aware at all times of the changes that have taken place and are taking place in the language. The resources of a language, and especially a language that has been the vehicle of a strong cultural development, are so vast that no individual can reasonably hope to master them in a lifetime. For these reasons alone it is inevitable that the translator will encounter difficulties from time to time. Not all these will be soluble in terms of grammar, but a large number certainly will. It is then that an explicit knowledge of grammatical features will stand him in good stead. Naturally he is not so much interested in grammar in general as in comparative grammar, the points on which the two languages concerned differ or agree, in particular the points of disagreement. It is not necessarily the unfamiliar feature that causes the most trouble. Often it is the familiar, that which is taken for granted, that represents the greatest difficulty.

On the simplest level, grammatical analysis is a description of the structures occurring in a typical text. The usual unit of description is the sentence. The constituent analysis of a sentence, a process akin to the parsing taught in school, would provide a diagram of a sentence structure on the following lines:

1a *Hans liebt die deutsche Sprache*
1b *Hans loves the German language*

S	P	
	P	O
Hans	liebt	die deutsche Sprache
Hans	loves	the German language

This is a straightforward sentence and one that presents no difficulty. The correspondence is virtually perfect, since the structure is S-P-O (NP + VP + NP) in both languages. But consider the following:

2 He sees a flying fish (D *einen fliegenden Fisch*)
3 He sees a writing desk (D *einen Schreibtisch*)

These sentences seem to be of similar structure to 1, and this is certainly true of the German version. The diagram would be similar to the first example (NP + VP + NP). Yet if we analyse the structure of the object more closely we see that German has the following:

NP			NP	
Art	A	N	Art	N
einen	fliegenden	Fisch	einen	Schreibtisch

and English has:

NP			NP		
Art	A	N	Art	A	N
a	flying	fish	a	writing	desk

Both 'flying' and 'writing' have the same verb stem + *-ing* form. Yet we know that they are different (though this is not as clear in English as it is in German). The diagram does not show this, although it is useful for gaining a general idea of the sentence structure. A paraphrase will clarify the matter:

2a He sees a fish
2b The fish flies

Similarly:

3a He sees a desk
3b The desk is for writing

This is further explained by the restriction on the subject of the verb *write* to animate creatures and, indeed, to human beings. Desks cannot write. But if the sentence were as follows:

4 He sees a flying desk

the explanation would not suffice.

4a He sees a desk
4b The desk flies

The verb *to fly* is not restricted to animate beings. This explains the ambiguity in the much-quoted sentence: 'Flying planes can be dangerous'.

We should have to add an explanation along the following lines: the fish flies by its own volition; the desk flies because somebody else propels it. In the first case the fish is an agent, in the second the desk is acted upon. Thus the fish makes itself fly, but somebody or something else makes the desk fly.

5 He saw a moving picture

This may mean:

5a He saw a picture
5b The picture moved
5c It was a film
5d It moved him

In translation we have to cope with ambiguities of this kind occurring in two languages at the same time, in the SL and in the TL. To understand the ambiguities and to avoid compounding them in the process of translating, we must dismantle the languages and look at the component parts.

The problem is complicated by the fact that we have to cope with two languages at once. A common basis for comparison has to be established; we need common grammatical categories. This is not an insuperable problem. The traditional terminology of grammar in modern languages is largely accidental in that it has been derived from the analyses made for Greek and Latin. Yet the grammatical structures of these languages differ from that of, say, English, and the classes and categories within them are not always appropriate to English. Grammar books often describe English as though it had the full range of tenses found in Latin. The Romance languages do have a series of tenses similar to Latin, but strictly speaking English has only the past and the non-past. Future and other forms are expressed by means of auxiliary verbs. To some extent this also applies to German. Similarly, there are difficulties in the use of classes such as noun, verb or adjective. The range of adverb and adjective in German does not always correspond with that in English.

There are, however, definite advantages in retaining the old terminology as far as possible. First of all, any other terminology will also lead to difficulties – the very newness of the terms will have to be explained and will not be entirely satisfactory. To establish a common basis for comparison will be a cumbersome and unrealistic process, since in practice the translator manages quite well without new terminologies. Providing the limitations and qualifications of the traditional terminology are borne in mind, it will serve. Further, and most important, a corrective is also supplied by the fact that we are not dependent exclusively on the grammatical level for our understanding of the transfer process, but on the higher levels, such as vocabulary and situation. The traditional terminology can be used with this in mind and will automatically provide common categories and classes that will create the working basis. It is easier to do this with languages such as German and English that are fairly closely related, but it would not be satisfactory with languages such as Japanese and English.

The obligatory-optional distinction is an indispensable one. One of the most important things to know, especially about the TL, is what *must* be said in it as against what *can* be said in it. It is vital to know that in modern English the auxiliary *do* must precede the subject noun and the main verb in certain types of interrogative: *Do you know?* rather than **Know you?*, which is normal in German (*Weisst du?* or *Wissen Sie?*).

A TL structure may sometimes appear to differ greatly from the SL structure, but it is possible that the TL has alternative forms that are nearer to the SL one while still conveying the same message. In technical literature, where conveying information is the main consideration, this point is highly relevant and therefore affects the criterion of adequacy. Generally one alternative is preferable to the others, if only by a slight margin. More often than not, the optional forms all tend to belong to a similar series (*have, possess*; or *supply, provide, furnish*); they are synonyms or near-synonyms, and only a deeper analysis will show that one of them is more appropriate for the particular utterance than the others. It is common, for example, for the alternative forms to belong to the same form class, as with noun and noun phrase, and for the divergence to be one of rank. The form class as such, N or NP, is obligatory. Where very divergent grammatical forms are possible (say, a choice between N or V forms), it may be that one introduces ambiguity while the other does not. It may be better to say: *He saw a film* or *He saw a moving picture at the cinema* than *He saw a moving picture*. It would certainly be wrong in this instance to say: *The picture he saw was moving*, but in another instance the change from N to V (participle) might be correct. The form may represent a considerable divergence from the original, or the substitution may take place on a level that cannot be called grammatical. A 'deep' analysis, if carried far enough, tends to take us into the lexical realm or even into the realm of situation.

The basic unit of grammar is the word. There are also larger units such as the word group or phrase and smaller units such as the morpheme. Morpheme and word may coincide, as in *bird*. Sometimes the morpheme is a stem (*find-* in German *finden*, *-fect* in English *effect, affect, confection, defection*) and the derivative elements (prefixes, suffixes).

These variations in the word are usually studied as the subject-matter of morphology, the branch of grammar that deals with the principles discernible in word-building processes. Each language has its own methods of forming new words from morphemes and of converting one part of speech into another. In English, adverbs are frequently formed by the addition of *-ly* to the end of an adjective (*that was a quick change, he changed quick-ly*). The Romance languages use a similar device (I *dolce*, sweet; *dolcemente*, sweet-*ly*; F *douce*, douce*ment*; S *dulce*, dulce*mente*). German has the suffix *-lich*, which also functions in this way (*schwer*, difficult; schwer*lich*, with difficulty; *wahr*, true; wahr*lich*, truly; *ledig*, single; ledig*lich*, solely).

The three principal word classes are nouns, verbs and adjectives. It has been estimated that half the vocabulary of German consists of nouns, a quarter of verbs and a sixth of the adjective-adverb group.[1] These are termed 'open classes', because they have a limitless ability to expand. The 'closed classes' (limited in number and potential for increase) or 'function classes' (because they serve to modify or relate members of the 'open classes' rather than to express an independent meaning) include prepositions, conjunctions, interjections, articles and pronouns. It is not always easy to define the latter. In German, certain types of adjective and adverb are grouped together; and the line between preposition and adverb is not absolutely clear cut. Demonstrative pronouns and articles are often called noun determiners, and a number of items are difficult to fit into any scheme of classification, such as the negators (*no, not*) and the affirmers (*yes*). Moreover the closed classes are not hermetically sealed off – as we have seen, the adverbs can be augmented by adding affixes to adjectives.

We need these terms to analyse and explain some of the processes that take place in the process of translation. Without them analysis would be extremely difficult. At this stage no better way has been found to define these familiar and in many ways unsatisfactory grammatical classes and categories.

The procedure known as transposition, for example, consists of replacing one word class in the SL with another in the TL, modifying categories, changing rank and so on. Here are some definitions:

categories	transitivity, intransitivity; tense; voice; modality; participle forms; degrees of comparison within adjectival forms; predicative and attributive use of adjectives; types of article (definite, indefinite, zero); negation; number; person
structures	co-ordination, modification
word classes	noun, verb, adjective, etc
syntactic classes	subject, object, predicate, etc
rank	morpheme, word, phrase, clause, sentence

I shall now give an outline of the main differences between word classes in German and English, occasionally adding examples of differences between other languages and English.

NOUNS

There is a correspondence between the behaviour of the German and the English noun over a wide range and on many levels. Many expressions are identical in form and to a considerable extent in meaning: *Hunger, Hand, Land*. Others are close enough in form (and have a sufficiently similar meaning range) to be recognizable without great difficulty: *Haus, house; Maus, mouse*. An advantage here is that a noun in written German is always recognizable as such by the fact that the initial letter is capitalized. In other languages only proper names have initial capitals (*John, Jean, Giovanni*).

A certain pattern can be detected in some similarities of this kind. In *Haus, house*, for instance, the letter *a* or the diphthong *au* appears in German where English has *o* (or *ou*). But this is not a universal rule and *die Faust* yields not **foust* but *fist*. Thus there is one-to-one correspondence on this level between *Haus* and *house*, but a one-to-none correspondence between *Faust* and *fist*. We could at a pinch say that there was a one-to-one correspondence between the noun *die Fäulnis* and *foulness*, but in fact the two nouns do not have the same range of meaning, though there is some over-lapping of denotation and connotation. (The normal equivalent of *Fäulnis* is *rottenness* or *putrefaction*.)

Similarly, the German letter *V* is often replaced by *f* in English, as in *Vater, father; Vorwort, foreword*. But in many words, especially those of foreign origin, this change does not take place: *Ventilation, ventilation; Variation, variation; Vegetation, vegetation*. There is a tendency for *t* within words to give *d* in English (*Garten, garden; Bett, bed*), but note that *Matte* corresponds to *mat* (not *mad*). The form *sch* is pronounced like *sh* in English and can often be replaced by it, as in *Schaf, sheep; Schärfung, sharpening*. Similarly, *k* corresponds to *c* (German having the hard *c* only in a small number of foreign loan words), as in *Kleie, clay; Karte, card*. These are merely rules of thumb that may sometimes be helpful.

To a large extent the German noun functions in a similar way to the English noun. The subject of a sentence and the object (if any) are usually noun forms in both languages. They may be preceded by an article. Only abstract nouns are not preceded in this way in English, as with *Beauty is skin deep*. There are parallels to this in German, such as *Tugend besteht, Schönheit vergeht* (virtue endures, beauty perishes). But there are also differences, as will be shown. The article may be accompanied by an adjective that occurs between the article and the noun in both languages (*der gute Mann, the good man*). In a language like Swedish the article may follow the noun, as in *dagen, the day* and *dagarna, the days*. In Italian and French the article precedes the noun but the adjective usually follows it (I *il uomo buono*, F *l'homme bon*).

The differences between the two languages are more marked when we turn to modifications for number. Both English and German nouns undergo variation. The singular-plural system of English is fairly simple and regular with the noun modifying as a rule by the addition of a final -*s* (*book, books*). The German procedure is rather less simple. Some nouns become plural by the addition of a final -*e*, as with *Tisch, Tische* (table, tables); some by adding final -*en*, as with *Frau, Frauen* (woman, women); and others by adding -*er*, as in *Brett, Bretter* (board, boards). But in some instances the plural is formed by internal modification, such as vowel change. This is occasionally seen in English, in forms such as *foot, feet*, but it is much more common in German, where a large category of nouns form the plural in this way, like *Mutter, Mütter* (mother, mothers), *Bruder, Brüder* (brothers, brothers).

It should be noted here that the definite article may indicate the plural in German, but it never does so in English. Thus the forms I have cited are usually given with the accompanying article: *der Tisch, die Tische; der Brett, die Bretter; der Bruder, die Brüder*. But there is no change for the nominative in the feminine gender and the singular has the same definite article as the plural (*die Frau, die Frauen*). As a result German sometimes has two indicators of the plural (the article form and the noun itself) but in some cases only one. In translating noun phrases from German to English a two-to-one correspondence is therefore common:

Art (def) (plural) + N (pl) = Art (def) (ø) + N (pl)

None of this applies to the indefinite article, for in both languages it is used only with singular noun forms and is omitted with the plural (*ein Tisch, a table* and *Tische, tables*).

1 Art (indef) (sing) + N (sing) = Art (indef) (sing) + N (sing)
2 ø + N (pl) = ø + N (pl).

In general uncountable nouns – which express the idea of a substance that is an undifferentiated mass, like snow, rain or water – tend to be parallel to one another in both languages. They do not normally exhibit any distinction between singular and plural. We speak of *Schnee, snow*, and *Wasser, water*, not of *ein Schnee, a snow* or *ein Wasser, a water*, except in special contexts. There is a good parallelism between the *Wein, wine*; *Haar, hair* type of noun, where the noun refers to a general class (as in 'wine is good to drink' or 'he has hair on his head'), but the plural form *Weine* and its English counterpart *wines* are not unusual ('the wines of France are very good') and the indefinite article is also possible ('this is a good wine'). For both languages the rule is that names of substances occur only in the singular when they represent a formless mass, as in *Milch, milk*; *Gold, gold*; *Leder, leather*; but when a distinct type is referred to, the plural can be used.

This use of the plural form to indicate a special type of a substance that is otherwise a formless mass is fairly common in technical language. The word for household salt is *Salz* (*das Salz*) in German and *salt* in English ('Pass me the salt'). In this context it is uncountable. But in chemistry we specify a special class of substances and use the plural *Salze* (*die Salze*) or *salts*, or the singular with an article.

6a Bei der Verdrängung einer Säure aus ihrem Salz durch eine andere Säure entsteht ein Salz dieser anderen Säure

6b The displacement of an acid from its salt by another acid results in a salt of this other acid

The uncountable noun *Salz, salt* thus becomes countable in the context of chemistry. This phenomenon is very common in technical language, which transforms the use of ordinary language items in specific ways. There are differences between German and English as regards denoting number in expressions for money, distances and so on. German, for instance, may have singular forms where English takes the plural:

zehn Mark	ten marks
fünf Yard	five yards
drei Mann	three men

But the contrast should also be noted between *drei Glas Bier*: *three glasses of beer* and *drei Tassen Kaffee*: *three cups of coffee*. In the first example the general pattern is followed but the second shows plurals in both languages. We have no space here to dwell on structural differences between the phrases in the two languages.

In both German and English, abstract nouns (and in technical texts these are more common than in the colloquial language) occur mostly in the singular (*Freiheit, freedom*; *Kälte, cold*; *Hitze, heat*), though they can occur in the plural in certain instances. In special cases we can have *eine Schönheit, a beauty* (a beautiful woman), where an abstract quality becomes personified, and the plural is therefore possible. Not all abstract nouns can be used in the same way, however. It is hard to think of a similar application for *heat* in English, though the German language allows us to say: *Ist das heute eine Hitze!* (*It's hot today!*).

One of the most important differences between German and English nouns relates to grammatical gender. The example of *der See* and *die See* has shown how gender can be used to indicate a distinction in meaning. Similarly, *der Moment* means *moment*, as in 'just one moment', while *das Moment* indicates *momentum* or *motive*. The gender of the German noun is thus important to its meaning. In English these distinctions are made in another way; usually, as in the above examples, another lexical form is used and no general rule for replacement can be given.

Gender in German is usually a grammatical device, whereas in English clear-cut distinctions of sex are made. In English a man is always *he* and a woman *she*. This is usually true in German too: *der Mann* and *die Frau* illustrate both grammatical gender and sex differentiation. But German also has *das Fräulein* for English *Miss*. A number of words in German have natural sex distinction (*der Stier, bull* and *die Kuh, cow*); but others (*die Forelle, trout*) are referred to simply as *it* in English. There are correspondences in words with sex derivations (*der Löwe, lion* and *die Löwin, lioness; der Hund, dog* and *die Hündin, bitch*).

Thus the German gender distinctions do sometimes correspond with the English distinctions of sex, but not always. There is in fact a certain arbitrariness in both languages. German is logical with *das Schiff* for *ship* (whereas in English ships are normally feminine). Germans speak of *die Tschechoslowakei* (Czechoslovakia), of *das Wallis* (the canton of Valais in Switzerland) but *der Sudan*. The seasons (*spring, summer, autumn, winter*) are neuter in English but masculine in German (*der Frühling, der Sommer, der Herbst, der Winter*).

In both German and English many nouns normally take the plural form. Sometimes the German form has a one-to-one correspondence in English (as in *Leute, people*), but the English may have a singular form as well as the plural, while the German occurs only as a plural (*Chemikalien, chemicals*) (but also a *chemical*).

Conversely, German has a singular form *die Hose* or plural *Hosen* where English has only *trousers*.

Generally other word classes, such as adjectives, can be nominalized in German (as they can in French, Italian or Spanish) by means of a preceding article. There is sometimes a one-to-one correspondence in English (*das Göttliche, the divine*) but certain English adjectives can be nominalized simply by using the word as a noun, whereas in the German counterparts the article is needed (*good, das Gute*; *right, das Recht*; *green, das Grün*). Again an infinitive form of the verb is nominalized in German by the addition of an article, as in *das Leben*, which can be translated into English either as a simple noun, *life*, or as an *-ing* form of the verb, *living*. Additionally, the infinitive form can sometimes be used without change (*to live*). An example of this is Pope's famous line: 'To err is human, to forgive divine.'

An important difference between German and English nouns is that the former can be declined by means of case inflections. Case indicates the relation of the object designated within the utterance or, to put it another way, the relation of the word to other words within the sentence. This relationship can be expressed either by a change in the word itself (by means of endings, prefixes or stress), or by word order. A noun generally has no more than four case endings in German. The declension of the word *Tisch* (table) yields only the forms *Tisch* (nominative and accusative singular), *Tisches* (genitive), *Tische* (dative singular; nominative, genitive and accusative plural). These are supplemented by the article, which is also declined in a similar way, so that we usually know what function the expression has in the sentence. In English only the word order gives us this information. Word order can perform this role in German too:

Die Frau singt (the woman sings)
Wir hören die Frau (we hear the woman).

But word order is not the sole indication of relationships in a German sentence as it is in English, and as a result there is a greater possibility of syntactic ambiguity in English than in German.

The basic contrast in German is between the nominative and the three other cases: genitive, dative and accusative. The nominative occurs as the subject of a sentence. The accusative is the usual case for the direct object, the object affected by the verb. The gentive can also be used as object, though this is becoming infrequent. It indicates possession, origin and that something is part of something. The dative marks the direction of a verbal action. It is used for the indirect object to whose advantage or disadvantage something happens. Case is an important category in German, but in English there is only one case ending, corresponding to the genitive and marked by *'s* at the end of the word (*the man's book*). In technical English even this is often dispensed with, so that *the engine's speed*, for example, is now replaced by *the engine speed*.

Only with the genitive, therefore, is there any likelihood of a one-to-one correspondence at this level. A German nominative or accusative is not in one-to-one correspondence with an English noun used as subject or object, since the German expression will contain more information than is given in the English. It is part of a system in which case endings are used and articles indicate relationships within a sentence. The English noun is part of a system in which word order is the main indication of sentence relation. Prepositions are sometimes employed in English to perform functions that would be performed by case endings in German. This is particularly true of the dative object, which can be expressed by word order in English, or by using a 'to + NP' form. Hence the sentence *Karl schenkt seinem Freund ein Buch* might be rendered as (a) *Karl gives his friend a book* or (b) *Karl gives a book to his friend*.

This suggests an amendment to the rule that German case is generally replaced in English by word order (one-to-nought correspondence), on the principle that German expressions that use case endings can sometimes be replaced in English by expressions that include a preposition.

The case ending then, indicates the syntactic role of the noun and acts as a guideline for the sentence as a whole. Article, adjective (if any) and noun in a phrase must conform in case, gender and number. The noun form must conform to the verb and the various other elements are regulated by case relations. This confirms the suggestion that the German sentence is packed with a greater amount of information than might be considered necessary in English. German is not economical, but it is less susceptible of misunderstanding.

VERBS

The German verb has personal endings (*ich liebe*, I love; *er liebt*, *he loves*), and also has a tendency to combine with personal pronouns and verbal auxiliary forms. The English verb system is simpler than the German system and very much simpler than that of languages like Italian and Spanish. Normally it displays only two personal endings: for the third person singular an -*s* is added to the base (e.g. *he loves* in contrast to *I/you/we/they love*). The German has four forms (*ich liebe*, *du liebst*, *Sie/sie/wir lieben*, *er/sie/es/ihr liebt*).

Only an exceptional English form, such as the verb *be*, exhibits a range of flexional change comparable with German (*I am, you are, he/she/it is, we/you/they are*), i.e. it has three forms.

A major classification of verbs in both German and English involves what are known as *full* and *auxiliary* verbs. The former are countless; the latter are limited in number. Both German and English have three primary auxiliaries (*sein, haben, werden*; *be, have, do*) and six modal auxiliaries (*sollen, wollen, können, mögen, müssen, dürfen*: *shall, will, can, may, must, ought*). There is, however, no one-to-one correspondence between any of these German and English verbs, all of which are in frequent use. A literal rendering of, say, *Er wird alter* would be *He becomes older*, but the proper equivalent is *He is getting older*. The German form is Pn + *werden* (3rd person sing present) + A, while the English version has the form Pn + *be* (3rd person sing present) + get (present part) + A. The German form is very important in that it is used for statements describing processes that cannot be expressed quite as directly and easily in English, and technological writing is very much concerned with processes. But the German sentence and the English are still close, especially if a less idiomatic version like *He is becoming older* is used.

By contrast, *werden* as an auxiliary no longer coincides on a one-to-one basis with *become*. It indicates the future, which corresponds to English *will* used as an auxiliary. The formula for expressing the future is Pn (personal) + *werden* + inf in German and Pn (personal) + *will* and inf in English. The translation of *ich werde lieben* is therefore *I will love* (and not *I become to love). These auxiliaries occur very frequently in technical texts, as indeed to all the auxiliaries, the primary ones in particular. In German *werden*, if used with an adjective or comparative form, can express the notion of a progressive change, augmentation of diminution:

76

7a *Das Feuer wird noch kräftiger* (literally: The fire is becoming still stronger)
7b *The fire is burning more fiercely*

Tun is solely a full verb in German, although it is related etymologically to English *do*. *Ich werde es tun* (Pn + Vaux + Pn + V inf) is translated as *I shall do it* (Pn + Vaux + V inf + Pn), which is a one-to-one correspondence on the word level, though not in word order. *Tun*, however, does not act like *do* to express emphasis, interrogation and so on. In fact *do* has no counterpart in such instances among the German auxiliary verbs.

As a linking form (linking subject and object), *sein* corresponds to *be*. There is a one-to-one correspondence between *Er ist der Mann* and *He is the man*, but when the same verb is used as auxiliary with full verbs, differences occur in function. That is, there are differences in the way in which it behaves in a compound verbal form. *Sein* + participle is used to express the past tense, as in *Ich bin gefahren*. A literal version of this would be **I am gone*.

Moreover, there are differences in range of tense between German and English. The range in both languages is, in any event, not as wide as that in Italian and Spanish. Time references other than past definite are achieved by means of auxiliary verbs, whereas the Romance languages make much more use of flexional change. In English only two forms express tense in this manner, the base form like *love* (*I/you/we love*) and the past, as in *loved* (*I/you/we loved*). Similarly German has *ich liebe* for the present and *ich liebte* for the past. Any other time references are expressed by an auxiliary plus the infinitive, whereas the Romance languages have inflected forms to express a wide range of tenses: Italian *amo* (present), *amerò* (future), *amai* (past) and so on. In German the future is expressed by Pn + *werden* + inf (*Sie werden lieben*) and in English usually by Pn + *will* + inf (*they will love*); the perfect by Pn + *haben/have* + past participle (*ich habe geliebt* and *I have loved*).

Haben mostly corresponds to English *have*. But it is in fact narrower in its range of application. It signifies only the condition of possessing something, whereas *have* may denote an event, such as *have a baby*, and may be causative, as in *have the house painted*. In the first example *have* corresponds to the German verb *bekommen* (*obtain, acquire*) and in the second it corresponds to *lassen*. It does

77

not fully match *haben* in its function as an auxiliary with past reference.

The modal auxiliaries are somewhat similar. Each pair overlaps to a certain extent, but they also differ over part of the range. Just as *haben*, resembling *have* in phonetic and written form, does not always correspond to it, so *sollen* appears to be like *shall* but is far from corresponding one-to-one to it, in many instances. It is a common error to equate the two. Yet *sollen* has an imperative application: in a phrase like *du sollst schweigen* a command is uttered, rather than a statement about a future happening. It is equivalent to *be silent*, rather than to **you shall be silent*, yet the latter type of mistranslation is often found in work by translators whose mother tongue is not English. Even those whose mother tongue is English sometimes slip into this error. *Sollen* is in addition used for indirect requests, as in *Er soll kommen*, which is equivalent to *he ought to come* (and not to **he shall come*).

At first sight *müssen* seems to be the counterpart of *must*. This is sometimes so, for *müssen* implies the most intense obligation, an objective and externally imposed compulsion. But the negative of *müssen* (*nicht müssen*) does not mean *must not* but rather *do not have to*. The expression *nicht dürfen* is the equivalent of *must not*. The sense of *dürfen* is that of moral obligation, duty, the desirability of doing something. It suggests an internal or psychological compulsion. It has much the same meaning as English *ought*.

The modal auxiliary *mögen* is used to express preference (*would like to*) as against *wollen*, which dignifies desire (*wish to*). The sense conveyed by *können* is of physical possibility and it is therefore equivalent to either *can* or *may*. In *er kann den Brief geschrieben haben*, the latter is appropriate: *he may have written the letter*. But in *wenn . . . angenommen wird, dass . . ., kann man auch den täglichen Bakterienzuwachs . . . errechnen*, it is best replaced by *can* (*Assuming . . . the daily increase in bacteria can be calculated*).

Akin to the problems of the modals are those encountered with the subjunctive. This form is very extensively used in the Romance languages and has very complex applications: it can express desire, possibility and grammatical subordination. The subjunctive is less common in German than in the Romance languages and certainly does not have the same range. Yet it is found more often in German than in English, where it has largely fallen out of use. Where we have to translate a subjunctive into English, nothing like a one-to-one correspondence can be expected. A modal auxiliary with an infinitive or a circumlocution will be the normal replacement. The subjunctive in German is used to express unreal conditions, desires, events that are not actually in existence. In English verbs like *may* or *were* approximate to the subjunctive function. The subjunctive form persists mainly in archaic, literary or ceremonial phrases. Examples are *God save the Queen, Heaven forbid that* and *if that were true* (hypothesis rejected) or *if only it were so* (unfulfilled wish).

A few remarks should be made here about German and English systems of indicating time. Tense is usually viewed as differentiating between action occurring in the present, the past or the future, and the nature of the action, its duration, its initiation and its completion are often described under the heading 'aspect'. Aspect is a category to which attention should be paid since it is a region where grammar and semantics meet.

In German the past refers only to past time and to nothing more, whereas English makes a distinction between the past definite, which refers to an action completed (*I saw him yesterday*), and the perfect, which deals with events begun in the past and continuing, or still having an effect, in the present (*I have written to him*), or conceived as timeless. Somewhat similar distinctions are made in Spanish. For example *fui al teatro la otra noche* (*I was at the theatre the other night*) refers to a remote past in relation to the present moment and to a process with a brief and precisely delineated development. The perfect, as in *he amado* (I loved), is less precisely outlined but indicates a more definite past. It is contrasted to the imperfect, *cuando le vi, él estaba en su casa* (*when I saw him, he was in his house*). But German and English do not exhibit this type of parallelism.

The tendency in German is for the present to predominate over the future:

8a   *Die Pralltellermühle entspricht den Anforderungen . . .*
8b   *The baffle plate disintegrator . . . will meet all demands*

The German sentence uses the present tense where the English version has a future reference; this is in fact optional, since the present could be used in the English, but the future is preferable. This tendency for German to have the present where English tends to use the future is common in other Germanic languages. But this example raises a more important issue. The 'timeless' type of construction (either *will meet all demands* or *meets all demands*) is very common in technical texts. Its most typical form is that of the statement about the properties of a substance: *water boils at* 100°*C*. This is equivalent to saying:

> *water always boils at* 100°C
> *water will always boil at* 100°C
> *water has always boiled at* 100°C

In this type of statement the present form – *water boils at* 100°*C* – is the usual and preferred one. It is the typical form of the scientific statement and the form in which scientific laws are expressed. It is the form of the equation translated into words:

$$\frac{m}{2} v^2 = eV$$

$$\frac{m}{2} v^2 \text{ is equal to eV}$$

The forms expressing time in German and English consist of simple forms such as the present (*lieben* and *love*) and the past definite (*liebte* and *loved*) and a series of compound forms consisting of auxiliary plus infinitive, or auxiliary plus participle. The correspondences here are only partial. German past and perfect forms (*liebte* and *haben geliebt*) may be the equivalent of English *loved, have loved, was loving* or *has been loving*. In other words the German forms have a wider range than the English ones (one-to-many correspondence) and the English forms conversely make distinctions that are not present in the German ones. Interference from German will generally result in these forms being used incorrectly. For instance a past definite *-d* form will be used instead of a *was + ing* construction. There is no similar form in German to relate a past event to a present result.

The compound forms are particularly troublesome. This can be seen if we compare the use of an auxiliary like *sein* as a full verb† with its use as an auxiliary in *sein* + participle. *Ich bin der Mann* translates as *I am the man*, but the correct equivalent of *ich bin gefahren* is not *I am gone. As previously pointed out, it should be translated as *I travelled*. In the case of *haben* (have) the situation is analogous. There is a one-to-one correspondence between *Ich habe es* and *I have it*, but a one-to-many correspondence between *Ich habe*

† The term 'lexical verb' is used by some writers on grammar instead of 'full verb'.

gegessen and *I have eaten, I ate*. Another source of confusion arises from the past definite forms. The expression *ich ass* can correspond to *I ate, I have eaten*. Hence *ich habe gegessen* shows a one-to-one correspondence with English, and *ich ass* exhibits the same correspondence. The completed past in German may require an adverbial qualifier, as in *ich habe schon gegessen* (literally, *I have already eaten*). A further case of one-to-two correspondence (on the word level) occurs with the present tense, *ich esse*, which corresponds either to *I am eating* or to *I eat*.

Related to this is the expression of the passive voice in the two languages. We normally contrast passive with active, according to whether the subject of the action is the performer (active) or the goal of the action (passive). There are, however, intermediate forms that frequently occur in languages like Spanish – the reflexive forms. The reflexive forms are generally translated into the passive in English. The passive occurs very frequently in technical language, and is in general more common in German than in English. The usual form for the passive is *sein* + part or *be* + part. The English passive thus formed is extremely susceptible to ambiguity because participles can function either as verbal elements or adjectivally. For example *he was wounded* may be paraphrased as *the man received wounds* or *the wounded man*. This is not so bad, but what of *the man was frying*? This could mean either that he was using a frying-pan to fry something or that he himself was being fried. This ambiguity is not found in German:

die Tür wird geschlossen (the door is being closed by someone)
die Tür ist geschlossen (it is a closed door)

Either of these could be translated into the ambiguous English: *the door is closed*.

The three non-finite verb forms – infinite, present participle and past participle – display a good deal of similarity in both languages. The German infinite always ends with the morpheme *-en* (or *-n*), as in *lieben* and *tun*. The English infinitive is the same as the base form, as in *love* and *do*, but it generally replaces the German in the form *to* + V, as in *to love* and *to do*. However, expressions in German of the V (full verb) type often have their English counterparts in a catenative, V (full) + *to* + V (full):

sie lernen schreiben
they learn to write

81

An alternative translation is the form V (full) + participle (*they study writing*).

Where the German infinitive is used as a nominal form (subject or object of a sentence) in the pattern V (infin) + V (fin) + A, the corresponding English expression will mostly be the same, with a *to* + V in the nominal function:

> *irren ist menschlich*
> *to err is human*

A German form similar to the English is sometimes called the 'prepositional infinitive'. This consists of *zu* + V (infin) and has a parallel in *to* + V (infin) in English:

> *es ist notwendig zu kommen*
> *it is necessary to come*

Again, in some cases a present participle can be used instead (though not in the example given above):

> *er ist fähig zu arbeiten*
> *he is capable of working*

The *zu* + V (infin) form is also used to express a modality (attitude of the speaker to his statement), usually necessity:

> *der Acker ist zu pflügen*
> *the field is to be ploughed*
> *das is zu prüfen*
> *this is to be tested (must be tested)*

The present participles display a high degree of correspondence and the *-end* ending, as in *liebend*, will normally find a counterpart in English *-ing*, as in *loving*. These participles can be used adjectivally (inflected like adjectives in German), as in *eine liebende Mutter* and *a loving mother*. They are also used in the *zu* + part form to express obligation or necessity in the passive, as in *der zu pflügende Acker* (literally: **the to-be-ploughed field*). This type of construction should generally be translated by an adjectival clause: *the field that is to be ploughed* or *the field to be ploughed*.

The past participle has the characteristic form in German of *ge-*†
+ verb stem + *(e)t* (*geredet* or *gelobt*) or *ge-* + verb stem + *(e)n* (*gesehen* or *getan*). The counterpart in English is verb stem + *(e)d*

† German past participles are mostly prefixed by *ge-*. Exceptions are verbs beginning with *be-, ent-, er-, ge-, ver-, zer-*, and foreign loan-words like *reduzieren*.

(*walked, hated*). Of course English also has forms with -*t* (*wept*), and irregular forms (*stood*). German makes a distinction between weak verbs, which follow the pattern described above, and strong verbs, which change the internal vowel (*brechen/gebrochen*). The past participle is also used in the formation of compound tenses (e.g. *haben* + part [past]). This tense corresponds to the English past definitive:

ich habe das Fahrzeug gefahren
I drove the vehicle

The form *sein* + part (past) again does not translate one-to-one into *be* + part (past), which is generally a passive form in English, but rather into *have* + part (past):

ich bin gefahren
I have gone or *I went*

The passive in German is mostly formed by *werden* + part (past):

Güteprüfungen ... werden auf der Baustelle durchgeführt
quality checks ... are carried out on the building site.

Certain impersonal passive forms cannot be reproduced in English except by circumlocution. An example is the *es* + *werde* + part (*past*) construction, as in *es wird getanzt* (literally: **it is being danced*), which is best rendered by *there is dancing* or *people are dancing*.

A further important type of expression is the reflexive, as in *die Tür öffnet sich leicht* (V [3rd person, present, sing] + Pn) (literally: **the door opens itself easily*), which is rendered in English by the passive: *the door is easily opened*. The equation thus: V + Pn = *be* + part (past). Incidentally this expression is ambiguous and can also be translated as *the door is slightly opened*.

Before leaving the verb one more important application in technical texts must be noted: the command or imperative form. This occurs in descriptions of experiments (*Add 1 ml H₂O to solution*), footnotes (*siehe auch die Bedingungen für den Nachweis der Eignung in Abschnitt 4.1.2* = *see also the conditions of compliance check in Section 4.1.2*) and instructions. The imperative form is that of the present and it is easy to see that it is related to the 'timeless' present (*The moon is made of green cheese*, now, in the past and for ever, and *Add sulphuric acid to water*, in order to achieve the desired result now, in the past and for ever).

In German the imperative usually has the ending -*e* (or this may be omitted): *siehe* (*see*), *öffne* (*open*), *bleib* (*stay*). There is also a plural form corresponding to the second person plural indicative: *arbeitet, helft*. In addition there is an impersonal form of command using the infinitive: *fahren* (*go*) at a traffic light, for instance, and the *zu* + part (present) form discussed above:

9a   *Eignungsprüfungen sind von einer hierfür amtlich anerkannten Prüfanstalt durchzuführen*

9b   *compliance tests must be carried out by a test laboratory officially recognized for the purpose*

ADJECTIVES

The German adjective is inflected to denote gender, number and case when it is used attributively, but does not change when its function is predicative. The English adjective is invariable whatever its function in the sentence. Therefore there is often a one-to-one grammatical correspondence in the second instance. The German construction *der Junge ist klein* has its counterpart in English (*the boy is small*); but the attributive form *der kleine Junge*, as against *the small boy*, represents a many-to-one grammatical correspondence.

The German expression conveys more information than the English, particularly if the phrase also contains an article that again denotes gender, number and case. This can be seen by means of a comparison of a typical declension of German adjectives in the singular – the plural will illustrate the same point – with the English equivalent.

Masculine

Nom	der kleine Junge	the small boy
Gen	des kleinen Jungen	(of) the small boy
Dat	dem kleinen Jungen	(to) the small boy
Acc	den kleinen Jungen	the small boy

Feminine

Nom	die gute Frau	the good woman
Gen	der guten Frau	(of) the good woman
Dat	der guten Frau	(to) the good woman
Acc	die gute Frau	the good woman

Neuter

Nom	das gute Kind	the good child
Gen	des guten Kindes	(of) the good child
Dat	dem guten Kind(e)	(to) the good child
Acc	das gute Kind	the good child

In many German grammatical descriptions adjective and adverb are classified together. The predicative adjective can be used to modify the verb in a sentence. It is then not inflected. Thus we have *der eifriger Junge* (*the enthusiastic boy*) in attributive application, but predicatively: *der Junge ist eifrig* (*the boy is enthusiastic*) and adverbially: *der Junge arbeitet eifrig*, which can be rendered as *the boy works enthusiastically*. Here German A/Adv + ø = English A + *-ly*. An extra element is required in English to convert the adjective into an adverb – the typical adverbializing suffix *-ly*, for we cannot say, as we can in German **the boy works enthusiastic*. A number of words in English do function as adverbs and adjectives: words like *fast* or *hard*; some words ending in *-ly*, like kindly, and others expressing periods of time (*hourly*, *weekly*, *monthly*). But in general German is in one-to-two grammatical correspondence with English. And while the German word is unaffected by its function, the English must in general have the *-ly* modification.

Degrees of comparison exhibit important differences of range. English stands midway between the Germanic and the Romance languages here and uses two types of comparison, a Germanic type and a Latin type. In German comparison of adjectives is expressed by adding the ending *-er* for the comparative and *-ste(n)* for the superlative. For example, *schnell* (*fast*), *schneller* (*faster*) and *der, die, das schnellste* (*fastest*). This parallels the English degrees of comparison formed by the endings *-er* and *-est* (*slow, slower, the slowest*). The German forms are modified to denote case, gender and number, just like the ordinary adjectives, so that we get, for example, *das schnelle, das schnellere, das schnellste*. But English also has its Latin forms on the lines of the Romance languages, in which the qualifiers *more* and *most* are used. This is like Italian *più, il più* and French *plus, le plus*, e.g. *più intelligente* (*more intelligent*), *il più intelligente* (*most intelligent*), *plus précis* (*more exact*), *le plus précis* (*most exact*).

In this area, too, German is in a general one-to-two relation to English, but this does not mean that there are always two alternatives from which to choose. We cannot say 'more fast' or 'beautifuller'; *die drei wichtigsten Zentrifugen-Bauarten* must be translated as *the three most important* (not *importantest) *centrifuge types*. *Etwas trockenerer Feststoff*, on the other hand, can only be rendered as *slightly drier* (not *more dry) *solid*.

The remaining word classes are members of the 'closed classes', which have a limited number of items.

DETERMINERS

The determiners include the articles, demonstratives and various other items such as *all* or *some*. In German (and in the Romance languages) the articles have a many-to-one correspondence with those in English, where this word class does not have the function of signalling categories like number, gender and (in German) case. The main functions of the article are to specify (to identify) and to classify. These functions correspond to the two types of article: definite and indefinite. The definite article before a noun identifies (singles out) a particular individual, *der Mann, the man* from among *Männer, men* in general. The indefinite article classifies the individual as one among others of the same kind, *ein Mann, a man*. In German the definite article not only performs this function but also conveys information that the noun is masculine, feminine or neuter: *der Versuch* (masc), *die Probe* (fem), *das Experiment* (neut). All of these can be translated as *the experiment*. Gender is important in German as a means of establishing relationships in a sentence (through concord) and because it indicates in certain instances which meaning of a noun is intended. For example *das Band* can be *the ribbon* but *der Band* is *the book*; *der Gehalt* is *the content* but *das Gehalt* is *the salary*; *der Kaffee* is a beverage (*coffee*) but *das Café* is *the café*, and so on. The article also indicates the case and thus tells us of the relationship between the noun and other elements in the sentence (subject, object, possession, direction and so on). In addition the definite article varies according to number, thus supplying information as to whether the accompanying noun is singular or plural. As we have seen, the German noun also generally varies according to number. The comparison between German and English would be something like this:

	D	E	
Art (def)	masc/fem/neut	ø	1:ø
	sing/pl	ø	1:ø
	nom/gen/dat/acc	ø	1:ø
Noun	masc/fem/neut	ø	1:ø
	sing/pl	sing/pl	1:1
	nom/gen/dat/acc	ø	1:ø

The indefinite article in German, as in English, has no plural form, but it does have gender and case. The plural form of Art (indef) (sing) + N (sing) in both languages is: Art (indef [ø] + N [pl]), or simply N (pl). The negator *kein*, which behaves like *ein* in the singular, does have a plural, resembling in this respect the possessive pronouns, *mein, mine; dein, yours*; and *sein, his/hers/its*. Thus *kein Mann* has a one-to-one correspondence with *no man*, but the plural *keine Männer* has a many-to-one correspondence with *no men*.

Because it specifies, the definite article does not precede a proper name that is already individual and specific. English therefore does not say **the James is here*. This is also true of German, except that the article has other functions besides specification, and is sometimes used where a preposition would suffice in English, as in *die Dramen des Sophokles, the plays of Sophocles*.

Variations occur in the use of the article, which is sometimes omitted in German where it is needed in English. Expressions like *Lehrer sein (to be a teacher)* are examples. It should be noted that this sort of article-less expression can be transformed into a sentence in which an article is used in German but not in English. Thus *Unterricht halten (to give instruction, to teach)* becomes *er hält einen guten Unterricht* (literally: **he holds a good instruction*).

The definite article singular in German can fuse with certain prepositions, as in *am (=an dem), zum (=zu dem), im (=in dem), ins (=in das), zur (=zu der)*. This can be translated into Pp + Art (def) in many instances, such as the following: *das liegt einerseits an der Beschränkung der Baulängen . . . andererseits am zusätzlichen Aufwand für die automatische Betriebsweise: this is partly due to limitations in the construction length . . . and on the other hand to the additional expense of the automatic mode of operation.* But because of the extremely wide range of meaning of prepositions, idiomatic usages are common here, and the Pp + Art pattern does not necessarily have a one-to-one correspondence, although a prepositional form is generally carried over. A common example of this is *zum Beispiel* (literally: **to the example*) which is always translated by *for example*.

DEMONSTRATIVE PRONOUNS

Dieser and *jener* correspond to English *this* and *that*. *Selbst* can be replaced by *self* forms (*Fritz selbst hat es gesagt: Fritz said it himself*). But the reflexive requires circumlocution: *Dies versteht sich von selbst* (literally: **this understands itself from itself*) corresponds to: *It is obvious* or *It is self-evident*. There is a greater parallelism between *er hat sich selbst für seine Idee geopfert* and *he sacrificed himself for his idea*. This form of *selbst* is to be distinguished from the adverbial use, which is generally equivalent to *even*.

Although *solcher* has more or less the same meaning as *such* in most contexts, it will be better in many cases to translate it simply as *this*. In a passage on centrifuges, the following clause might occur: *es gibt jedoch auch solche, die beide Arbeitsprinzipien kombinieren* (literally: **there are however also such that combine the two operating principles*). This translates as: *there are, however, also centrifuges that combine both principles* (of operation). Similarly, *Ich habe solchen Hunger* should be rendered as *I am so hungry*, but *solches Wetter* may become *such weather*. A form like *ein + solche* is not of course literally reproducible and *ein solches Wetter* must become simply *such weather* or *weather like this*, and *ein solch Gefühl, a feeling like this*.

Indefinite pronouns differ in their application in many instances and also differ in their structural relationships to some extent. The German *all* agrees with the English use of *all* in *alle Bäume, all trees*, but it has many idiomatic applications that are not the same, such as *alle Welt* (*everyone*). It also occurs after personal pronouns, as in *sie alle, all of them*, and after demonstratives like *das alles, all this*. Also, the English equivalent of *all* may be *everything*, as in *alles rennt, everything races* (*runs*).

Beide has the sense of 'the two' rather than of 'both' – *die beiden Mädchen, the two girls*. The word *etwas* is generally equivalent to *something* – *etwas Neues, something new* – but it can also be rendered as *some*, as in *gib mir etwas davon, give me some of that*, and *somewhat* in *etwas salzig, somewhat* (*rather*) *salty*.

The problem with *jeder* is to decide whether it is to be rendered as *every* or *each*.

Of special interest is the negating pronoun *kein*, which can be rendered as *none, no* or *not*. This is dealt with further under syntactical range.

NUMERALS

There is considerable coincidence between English and German in the use of numerals. The correspondences are largely on the lexical level and numerals also correspond one-to-one on the grammatical level (*null = zero, eins = one, zwei = two*). The only points of difference are that in German these numerals inflect. The numbers after twenty have a different construction from their English equivalents (*einundzwanzig* [literally: *oneandtwenty], *twenty one*; *hunderteins* [literally: *hundredone], *a hundred and one*. Dates in German are expressed as follows: 10. Oktober 1969. Time expressions are as below:

es ist eins	*it is one (o'clock)*
es ist ein Uhr	*it is one o'clock*
fünf Minuten vor zwölf	*five to twelve* (literally: *five minutes before twelve)
es ist Viertel nach neun	*it is a quarter past nine*
dreiviertel zehn (Berlin) ⎫	
viertel vor zehn (Rhineland) ⎭	*a quarter to ten*

The ordinal numbers also show a close correspondence between the two languages: *der erste, the first; der zweite, the second*. The series *erstens, zweitens* also corresponds to *firstly, secondly*.

In fractions the suffix *-tel* corresponds to *-th* in English.

zwei Fünftel	*two-fifths*

but

Hälfte	*half*
ein Drittel	*a third*

German has no preposition of possession in constructions like *mit einem achtel Kilo = with an eighth of a kilo*.

The affix *-fach* generally corresponds to *-fold*, as in *zweifach = twofold* (*twice*), but *mehrfach* may be *several* rather than *manifold*, and the first in the series, *einfach*, cannot be rendered as *onefold; it is usually translated as *simple* or *simply*. The suffix *-mal* corresponds with *-ce* in *einmal, once; zweimal, twice; dreimal, thrice*; but from *viermal* onwards the expression *times* must be used (*four times*). These expressions may take the ending *-ig* and are then used adjectivally, as in *einmalig*. This adjective is difficult to reproduce in English (*non-recurring, unique*). That is, the answer lies on the lexical level. The grammar book does not help; the dictionary must be consulted.

PRONOUNS

The personal pronouns do not as a rule represent a source of particular difficulty in translating from German into English. German pronouns are inflected to denote case, number and gender, while English pronouns have a more limited use of inflection to denote case (*I, me; he, him*) and number (*I, we; he, they*) and also have forms that correspond to gender in the third person singular (*er, he; sie, she; es, it*). Forms of address differ, as English does not distinguish between familiar and polite *you* (except in archaic contexts, where *thou* may occur) as German does – *Sie*, polite, singular and plural; *du*, familiar singular; *ihr*, familiar plural. This distinction is also found in some of the Romance languages: Italian *Lei* (sing) and *loro* (pl) for polite, *tu* (sing) and *voi* (pl) for familiar; Spanish *Usted, Ustedes* polite, *tù* and *vosotros* familiar.

ADVERBS

As we have seen, adverbs and predicative adjectives are indistinguishable in German. The adverbial form in English is generally one that ends with suffix -ly, and the addition of this morpheme is a common way of transforming an adjective into an adverb. In German a word like *befriedigend* can be used adjectivally, as in marking school work: *sehr gut – gut befriedigend – ausreichend – mangelhaft – ungenügend* (i.e. ranging from very good to inadequate), where it corresponds to English *satisfactory* (*the work is satisfactory*). It can also be used in the same form adverbially: *die Anlage arbeitet bereits über zehn Jahre befriedigend* = *the plant has operated satisfactorily for more than ten years*. Again we can say *gute Leute* (*good people*), *diese Leute sind gut* (*these people are good*) and *diese Leute arbeiten gut* (*these people work well*). The correspondence in examples of this kind is nought-to-one.

The German adverb can also impose a time qualification on the verb, which in English has to be rendered by means of progressive tenses or catenatives: . . . *steigt die Mürbigkeit auch nach 30 Stunden noch weiter an . . .* = . . . *mellowness continues to increase even after 30 hours* . . . (increase – continue to increase); (literally: *the mellowness rises also still further after 30 hours*).

As in English, modality may be expressed in German by adverbs, although the usage is not likely to be parallel:

10a *Es gibt in der ganzen Welt wohl nicht zwei Mühlen, die völlig gleich arbeiten*
(literally: *there are in the whole world indeed not two mills that function fully alike*)

10b *it is doubtful whether there are two mills in the whole world operating with the same methods*

The use of modal adverbs in this way is much more common in German than in English and the above example is typical.

A peculiarity of German is a group of adverbs that can be transformed into attributive adjectives by means of the ending *-ig*. Examples are *jetzt*, (*now*), *jetzig*; *heute* (*today*), *heutig*. These attributive adjectives have no parallel in English. *Das jetzige Theater* (literally: *the now theatre) can be translated as *the theatre at present* or *the theatre now*, and *der heutige Tag* (literally: *the today day) as *the present time*.

Adverbs and adverbialized words are in general used more in German than in English. The capacity of German to adverbialize and to use adverbial constructions where this would not be possible in English is exemplified in the prohibition: *polizeilich Verboten* (literally: **policely prohibited*), *prohibited by police order*. Thus *Rauchen polizeilich verboten* = *no smoking*.

Another feature is the frequent use of elements to intensify or attenuate adjectives or adverbs in the superlative. The adverbs of degree include words like *ganz* (*entirely*), *äusserst* (*extremely*), *überaus* (*exceedingly*) and so on.

PREPOSITIONS

The preposition is sometimes called *Verhältniswort* (relation word) in German. It relates one noun to another, a noun to a verb or a noun to an adjective or adverb. Prepositions are also used in English to express functions that would be expressed in German by case endings. Indeed they are used more for this purpose in English than they are in German. English, for example, makes greater use of the preposition *of* to indicate possession than German does of the preposition *von*, which is replaced in German by the genitive case. Similarly the German dative is often replaced in English by the preposition *to*.

Prepositions in German govern particular cases (i.e. a specific preposition must be followed by a specific case). This does not happen in English, which has nothing comparable to the Spanish manner of distinguishing personal complements from inanimate ones by means of *a* as in *veo a Pedro, I see Pedro* and *conozco Madrid, I know Madrid*.

In the Romance languages, too, prepositions are used to link elements of a technical term that are simply juxtaposed in English:

I   *ciclo di sollecitazione*   *stress cycle*
F   *l'élément en combustion*   *the burning element*
S   *fuente de alimentación*   *power pack*

This N + Pp + N pattern is undoubtedly the most common one in Romance technical terminology. German tends to use juxtaposition (N + N) or fusion of a qualifying element with the noun.

The range of application of prepositions is so wide that it is impossible to provide a brief set of rules for translation and they must be dealt with as lexical items.

CONJUNCTIONS

Conjunctions in German are classified according to their effect on the position of the verb in the sentence that follows. Co-ordinating conjunctions have no effect, while subordinating conjunctions require an inversion of subject and verb. The co-ordinating conjunctions include words that have a one-to-one correspondence with English (*und, and*). These generally cause little difficulty for the translator. The pattern is NP + NP or sentence + sentence in both languages.

Some conjunctions (like *und*) can be used only to co-ordinate; some only to subordinate; and others for either function. *Wenn (if)*, for example, is used only to subordinate, and forms sentences on the lines: NP + VP + Cn + VP + NP, which must be translated as NP + VP + Cn + NP + VP:

> *wenn der Rechercheur seine Neuheitsvergleiche bei geringstem Zeitaufwand ausführen will, hat er zu berücksichtigen, das* ...
> *if the researcher wishes to carry out his novelty comparisons in as short a time as possible, he must remember that* ...

N+N co-ordinations, i.e. of two words or phrases, do not affect word order (e.g. *er ruft oder er winkt uns, he is calling or beckoning us*).

CHAPTER 7

Word formation

಄಄಄಄಄಄಄

EVERY language has methods for expanding its vocabulary from an existing stock, in addition to its borrowings from other languages. The languages of Western Europe employ two principal methods for extending the applications of indigenous linguistic elements.

New words can be formed by combining two independent words to create a new compound that functions as a single word. Sometimes the words are not actually fused but become associated with one another and are to all intents and purposes a lexical unit, behaving like one word. *Atom* and *Bombe* are separate words, but in Germany they can be joined together to create a new word, *Atombombe*. This can also be done in English, as where *black* and *board* have merged into *blackboard*. But in general English words do not fuse so readily as German words; they tend to remain apart even when they have been associated for a long time as one unit (*guinea pig*) or are linked by a hyphen (*free-standing*).

A morpheme may be added in front of a word (prefixation) or after it (suffixation). If we symbolize a word by the capital letter W and a morpheme by the lower-case letter w, the main possibilities for word creation by compounding (word and word) and derivation (word and morpheme) are as follows:

1 WW   *Atombombe, Quantenphysik, photoelectric*
2 W/w + W   *fester Körper (Physik)*†
3 Ww   *kindlich, childish*
4 wW   *berechnen (calculate, presuppose)*

More complex combinations may be based on the above, such as wWW, WWw, wWw, Www, W + Ww and so on. The elements of the combinations may be N + N, A + N, w + N, w + V, V + w, N + w, A + w and many others.

† Note, however, the form *Festkörper* (Physik), solid state (physics).

There are many similarities between German and English methods of word creation, and on the whole they are closer to one another in this respect than to the Romance languages. But two important points must be borne in mind: firstly, there is not always an exact parallelism; secondly, the compound (even a loose combination as in the second type) is not inevitably the sum of its parts. The German word *Handschuh*, for instance, is formed from *Hand* (*hand*) and *Schuh* (*shoe*) and should therefore be **hand-shoe* in English; but it is in practice translated as *glove*.

A similar observation can be made with regard to the process known as derivation. Many of the prefixes and suffixes used in German and English have a superficial resemblance and overlap in range of application to some extent, but the correspondence is usually incomplete. Indeed reflection will show that even in one and the same language there is often no general rule regarding the meaning of an affix. The word *jailer* seems to stand in the same relation to *jail* as *prisoner* to *prison*, but in fact a jailer is a person who supervises a jail, whereas a prisoner is an inmate of a prison. The same irregularity can be seen in *red*: *reddish* (meaning fairly red) and *girl*: *girlish* (meaning like a girl).

One would expect the formation of compounds to be more consistent in technical vocabularies than in ordinary vocabularies, yet only in certain fields is this type of regularity found. It is a desideratum that is more difficult to achieve than it appears at first sight, and it is fallacious to assume, as many people do, that technical language has completely overcome the untidiness of the general language.

A compound consists of at least two elements, both originally dependent words; one now becomes the main element (W) and the other a qualifying or determinant element (w), regardless of word classes. Thus it is not only in an adjective + noun combination that the first becomes a determinant element; the rule also applies to two nouns. In *Atombombe*, *Atom* has become the determinant element (w) and *Bombe* the main element (W). A *blackboard* is a type of board, and we write *blackboards* not **blacksboard*. Where one of the elements is an adjective, the determinant:main element relation is clear. For example we have no difficulty in recognizing that *Gleichstrom* (*direct current*) consists of (w) *gleich* (adjective) and (W) *Strom* (noun) and is in opposition to *Wechselstrom* = (w) *Wechsel* (noun) + (W) *Strom* (noun) (*alternating current*). We note

too that the German compound A + N = English A + N and
N + N = A (gerund) + N.

A technical term such as *Gleichstromnebenschlussmotor* is a typical
German jawbreaker, though not the most formidable of its kind.
and the formula outlined above can be applied to it just as well as
to simpler items. The basic pattern is the W + W type, but it is
also of type 4. For a start it can be subdivided into three elements,
of which two are themselves combinations: *Gleichstrom* + *Neben-
schluss* + *Motor* (wwW). The determining elements are of the wW
and wW type. The main element *Motor* can be rendered in English
as *motor* in this context. In many other contexts it would have to be
translated as engine, but here there is a full correspondence. *Neben-
schluss* (wW) is composed of *neben* (*by the side of*) and *Schluss* (*shutting,
closing contact*); its English terminological equivalent is usually
shunt-wound (WW). The German word is a preposition + noun
compound, the English a noun + past participle compound. Both
Gleichstrom and *direct current* are adjective + noun compounds. The
full translation is *direct current shunt-wound motor*. The German is
basically of type 1, the English of type 2. The comparison WWW
(wWwWW) with W + W + W (W+W + WW + W) shows the
divergence in structure between the two expressions.

Compounds have a hierarchical structure and can be viewed as
replacements for longer phrases or even sentences. In fact type 2
combinations behave like single lexical units, i.e. like words. But
these phrases may themselves be an abbreviated version of a longer
expression. The *direct current shunt-wound motor* is 'an electric motor
that has a field winding connected in shunt across the armature
circuit' and that operates on direct current. The first determinant
is the *shunt-wound*, which qualifies *motor* and is itself further qualified by the second determinant, *direct current*. The expression is
shorthand for the following: 'This is a motor. It is shunt-wound.
It operates on direct current' (object – type of construction – power
source). The German word can be analysed in the same way, as can
similar expressions following this complex structure. For example
Hochvakuumaufdampfanlage is also of the wwW pattern (divisible
further into wW + wW + W). The main element is *Anlage* (plant,
installation), with the first determiner *aufdampf-* (coating) and the
second determiner *Hochvakuum* (high vacuum). Hence a plant, a
coating-plant, a high-vacuum coating plant (w+W + w+W or
W + W).

From these examples a general rule for translating German compounds of type 1 and type 2 emerges. German runs the elements into one another and creates a single word, whereas English tends to keep them separate and at most will use a hyphen as a linking element.

Separation of elements is relatively simple when the combination is one that preserves the meaning of the original components, as in *Atombombe*, or where it can be inferred, as in *Handschuh*. The procedure then involves moving from WW to W + W (type 1 to type 2 or possibly from type 3 or 4 to type 2). But in many expressions the meaning of the elements is submerged in the compound. For example *baumlang* (adjectival main element qualified by a noun determinant) cannot be rendered as *treelong or *treetall; it must be translated as *lanky*. It is an idiomatic expression. Idiomatic expressions are not usually logical in structure, as this one is, for they often contain obsolete beliefs and ideas. It would be impossible to analyse a unit like *guinea pig* logically, for example, although it is an expression much used in technical English. The concept is of something that is not a pig, that does not cost a guinea, that does not come from the coast of Guinea. One might even say that it is not even usually a guinea pig, since it may be a rabbit or even a human being.

Derivation is a type 3 or type 4 process by which a word can move from one word class to another. It is one of the most productive ways of increasing a vocabulary. The process of prefixation (wW) usually results in a shift of meaning, while the addition of suffixes (Ww) will in addition entail a change of word class (certain prefixes can also have this effect). The addition of certain affixes can make a word into an agent (e.g. of the *-tor* type, as in *navigator*); can make it abstract (e.g. *-ation*, as in *navigation*); strengthen (e.g. *ur-* in *uralt*, making *old* into *primeval*); diminish (*-chen* in *Bändchen*, making *Band*, *volume*, into *small book* or *booklet*; and *Teilchen*, making *Teil*, *part*, into *particle*). A verb may become inchoative (expressing the start of an action) if it is given the prefix *er-*, or iterative (expressing repeated action) if it is given the suffix *-eln*.

There is a good deal of parallelism between the derivation processes of German and English, but unfortunately the function of a morpheme is not normally regular and the correspondence is not consistent, though it is not entirely haphazard either.

The example of *-lich* has already been given. It is sometimes similar in effect to English *-like*. Thus *kindlich* can be translated as *childlike*. But *weiblich* corresponds to *feminine*, not **womanlike*, though *womanly* is a possible equivalent.

Derivations may be classified in terms of the changes in word class:

a Verb-forming type: noun, adjective or verb transformed into a verb (the last is the conversion of a verb into another type of verb)
b nominalizing type: verb or adjective transformed into noun
c adjective-forming type: verb or noun becomes adjective.

As with the verb-forming type, a noun may be changed into a different type of noun by the addition of a suffix or prefix, and an adjective may also change within its word class in this way.

In German the main verb-forming suffixes are *-en*, *-eln*, *-ern* and *-ieren*. In general these do not correspond exactly with English forms, since English can use the process of conversion much more freely than German. In English it is possible to move easily from one word class to another without actually changing the word. The word is simply given a different function. This cannot be done to the same extent in German, where use of the class-changing morphemes is more common. If we wish to transform the German noun *Rat* (counsel) into a verb, we must add *-en* on to the end and produce *raten* (*to counsel*), whereas in English there is no change in form at all. Similarly to obtain a verb from *Rad* (*wheel, bicycle*), the suffix *-eln* must be added, while in English *cycle* or *bicycle* can be used as a verb without undergoing any change.

Some of the suffixes used for verb-formation in German were originally foreign borrowings, mostly from the classical languages; here there is a greater likelihood of correspondence, as the same borrowings have been made in other languages, including English. A case in point is the suffix *-ieren*, which is used to turn loan words from other languages (and also some indigenous words) into verbs. To some extent this is the counterpart of English *-ate*. The following list provides an illustration.

detonieren	*detonate* (Latin *detonare*: 'to thunder')
fabrizieren	*fabricate*
simulieren	*simulate*
ventilieren	*ventilate*

But we also have:

montieren	*assemble* (F *monter*)
reparieren	*repair*
zementieren	*cement*

Prefixes play a more important role in forming new German verbs from existing verbs. The morpheme *be-*, for instance, can be described as a transitivizing prefix. It tends to convert verbs derived from nouns into verbs that act upon an object. The formula is wWw. The noun *Last* (load), for example, gives rise to *be* + *last* + *en* = *belasten* (*to load*). This process is similar to some applications of the English prefix *be-*, as in *bemoan, befoul, befriend*, which are also transitivized. But such verbs are somewhat archaic.

German is particularly rich in prefixes and suffixes and in its range of compositions. While English also has an abundance of these the counterpart to a German prefix may well be a prepositional form.

An important feature of the prefixation of verbs in German (which more often than not is replaced by suffixation, or by a preposition after the verb, in English) is that it can endow words in the everyday language with a special technical meaning. The German use of prefixes is generally more precise than that of English (the same is true of compounding), which is one reason for difficulty in translation.

In addition to these morphemes the German verb can also be combined with a host of prepositions that precede the verb, while their English counterpart will follow it. In German the prepositions tend to fuse with the verb. These are type 1 (WW) word + word compounds that are very close to type 4 (wW). They are of type 1 where the sense of the preposition is still preserved in the compound. Where the sense has become submerged they have an affinity with the prefix and word combinations. The meaning of *durch* (*through*) in *durch* + V forms is often clearly retained: *durchfliessen, to flow through*. Because of the frequency with which Latin-derived words are used in English, especially in technical English, the parallelism may be less evident, as with *durchdringen, to penetrate*. Style may also play a part here. An English equivalent closer to the German may be unsuitable for a formal text, and technical prose is more often than not formal. Thus *durchschneiden* could in some circumstances be translated as *to cut through*, which is the more literal equivalent, but *to intersect* is likely to be a better rendering.

The same is true of other forms. The English counterpart for *über* is over, so *überblicken* can be translated as *to look over*. But it also corresponds to *to survey*. We must note here, incidentally, that where a more literal equivalent is possible, the equation is Pp + V = V + Pp. In English we must reverse the German order. *Overlook* may have the opposite sense (*ignore*) to *look over*, while *to through flow* and *to through freeze* are ungrammatical. But in any event the parallelism breaks down with words like *überlegen*, which can correspond to *to reflect, to consider* when the verb is used transitively, but which means *to be superior to* when the original is used intransitively. Note also the following:

um-	(round) + V	umfassen	embrace	(fassen: grasp, seize)
unter-	(under)	unternehmen	undertake	(nehmen: take)
	(among)	unterscheiden	distinguish	(scheiden: separate)
wider-	(contrary to)	widersprechen	contradict	(sprechen: speak)

With some ingenuity it is at times possible to deduce an accurate meaning from the elements that make up the verb: e.g. *umfassen*, grasp or seize round; *umhüllen*, wrap round; *unterscheiden*, separate among; *widersprechen*, speak contrary to.

Nouns form the largest class of words and also the basis of the largest number of combinations. For example it has been found that among compounds in agricultural terminology type 1 (NN) compounds comprise 46 per cent of the total. There is a large number of derivational forms by which nouns may be created from verbs, adjectives and other nouns. Two of the suffixes most commonly used for nominalizing verbs in German are *-ung* and *-nis*. Etymologically these are the equivalents of English *-ing* and *-ness*. Examples of the *-ung* series are words like *Kühlung*, which describes a continuous action like *cooling*, but also applies to more intensive *refrigeration*; an activity and its conclusion (*Prüfung, examination*); the result of an activity (*Zeichnung, drawing*) and the means of an activity (*Wohnung, residence*). The Latinized equivalents in *-ation* are fairly common, but not so common as to justify a general principle. The forms derived from the *-ieren* (*-ize*) type of verbs compare better (*Magnetisierung, magnetization; Neutralisierung, neutralization; Galvanisierung, galvanization*). These are all international terms that are characteristic of technical language.

Nominalized forms of verbs tend to occur with greater frequency in technical language than in ordinary speech. This tendency is expressed not only in the proliferation of Ww forms but also of type 2, w+Ww, in phrasal units. It has been pointed out that nowadays instead of *um Stahl zu gewinnen* (*to obtain steel*), we are more likely to have *zur Stahl + gewinnung* (*for the production of steel*); similarly instead of (*Neigung*) *Rost zu bilden* ([*tendency*] *to form rust*), (*Neigung*) *zur Bildung von Rost* ([*tendency*] *to formation of rust*) is more common. This conforms with the development in technical English by which a high degree of nominalization takes place. To a certain extent this facilitates translation, since it brings the German constructions closer to the English. The effect is one of impersonality, timelessness and objectivity. No actor, time, gender, mood or number are indicated. The disadvantage is the creation of a variety of expressions with subtle distinctions that are difficult to grasp and make the discovery of equivalents in a TL more difficult.

Forms ending in -*nis* are less common and also show less correspondence with English, despite the resemblance in shape and function to -*ness*:

Erzeugnis	*product* (*erzeugen, to produce*)
Ergebnis	*result* (*ergeben, to produce, to yield*)
Kenntnis, Kenntnisse	*knowledge, information* (*kennen, to know*)
Verzeichnis	*index* (*verzeichen, to note, to record*)

The -*nis* suffix, as these examples show, tends to denote the result of an action. The fact that the literal reproduction of a German expression often results in a colloquial or archaic English word is a useful guide to translation. If we approach *Wohnung* in this way, for example, we proceed from the verb *wohnen, to dwell*, to *dwelling*. If we substitute for this casual style of utterance a more formal one, we obtain *residence*. If we know the verb base of a -*nis* form we can proceed similarly. The result of the action of producing (*erzeugen*) is *a product*. (This would not assist, of course, with *verzeichen*, which displays polysemy, and the meaning would therefore have to be determined from the context.) *Kenntnis* can mean *acquaintance*, someone who is known, but also some*thing* that is known. In a technical utterance the latter is the more likely interpretation.

German verbs can also be used as nouns without modification if we add an article before the infinitive. This process, as we have seen, is far more common in English. The equivalent in English is usually the *-ing* form of the verb: *das Schweissen (schweissen: to weld)* = *welding*. Though this infinitive form is the one preferred by language purists, there is an increasing tendency in German technical writing to use the *-ung* form (*die Schweissung*). The remarks made above concerning the justification for this also apply here: the translation correspondence is generally closer, but there is no denying that it leads to duplication and pompousness.

A very large group of nouns is derived from verbs by the addition of the suffix *-er*. Both in German and in English the suffix is related to the Latin agential prefix *-arius*, which was used to designate one's occupation. This is still one of its functions in modern languages, but the range of application has been extended to include equipment and devices:

Dreher	*turner*	(*drehen*: *to turn*)
Heizer	*stoker*	(*heizen*: *to heat*)
Schweisser	*welder*	(*schweissen*: *to weld*)

The two languages also have a whole range of words ending in *-werker* (*werk* + *er*) or *-worker* (*work* + *-er*), such as *Stahlwerker, steelworker*.

The group of nouns ending in *-er* that designate a device or tool is on the increase. Examples are words like *Bohrer*, derived from *bohren* and meaning *borer* or *drill*; *Regler* from *regeln* and meaning *regulator*; *Senker* (*countersink bit*), from *senken*, *to sink* (something) and to *countersink*. Again the German *-er* does not necessarily correspond to English *-er* or *-ator*. The English expression, as with the *-tor* (*-ator*) forms, is frequently drawn from the classical languages, whereas the German is built up from indigenous elements. We often find, of course, a classically derived expression side by side with the Germanic one. Thus *Hubschrauber* is literally a lifting screw, whereas *helicopter* (which also exists in German) is built up from the Greek elements *helix* (screw) and *pteron* (wing). A *Bildwerfer* is something that literally throws a picture and a *projector* projects it (from Latin *projicere, to throw before*). A *Ventilator* is also a *Lüfter* (*airer*), which again corresponds to English *ventilator* (from Latin *ventilare, to blow*). There is a high probability that a German expression that is fairly easily recognizable from its elements has a Latinized counterpart in English. Naturally there is greater similarity between Romance-language terms and English terms. Almost any list of technical terms will tend to show affinity between the Romance group and English, while German is generally the odd man out:

E	F	I	S
helicopter	*hélicoptère*	*elicottero*	*helicóptero*
projector	*projecteur*	*proiettore*	*proyector*
ventilator	*ventilateur*	*ventilatore*	*ventilador*

In German, words ending in *-er* sometimes replace longer terms for devices. German comes closer here to English practice. For example *Frachtschiff* is the German for English *cargo boat* or *freighter*. *Frachtschiff* and *cargo boat* are the more usual type of construction for German. English, however, has a strong predilection for more compact words like *freighter* (freight ship). The tendency in German to replace a word like *Bananenfrachtschiff* by *Bananenfrachter* (*banana boat, banana freighter*) is thus a move in the direction of greater precision and at the same time of increased brevity. The same process is at work when *Mischmaschine* (*mixing machine*) is replaced by *Betonmischer* (*concrete mixer*).

A large group of German nouns is formed from adjectives by the addition of the suffixes *-heit*, *-keit* or *-igkeit*. Etymologically the English equivalent is *-hood*, but in practice it will generally be *-ness* or *-ity*:

Blindheit	*blindness*
Feinheit	*fineness*
Festigkeit	*firmness, solidity*
Gesamtheit	*totality*
Helligkeit	*clearness, brightness*
Möglichkeit	*possibility*
Wahrscheinlichkeit	*probability* (but also *likelihood*)

Whereas the *-ung*, *-ation* series tends to be terminations for abstract nouns and to indicate actions or processes, the *-er* endings designate agents (human beings, devices) and the *-heit* suffixes mostly indicate qualities.

A large group of terminations in German properly belong to type 1 (WW) because such terminations can function as independent words, but are so common especially in the technical language, that they can be listed in much the same way as the suffixes. Examples are *Stärke* (*strength*) and *Fähigkeit*, *capability*, itself a type 3

(Ww) derivation of the adjective *fähig* (*capable of*) + *-heit*. From *Beleuchtung* (*illumination*), which is obtained from *beleuchten* (transitivizing prefix *be-* + *Leucht* [light] + verb-forming termination *-en*), to *illuminate*, we get *Beleuchtungstärke* (literally: *strength of illumination*), from a type 1 compound, *Beleuchtung* + *Stärke*. This means *illumination, screen brightness*. The English equivalents are type 3 (Ww) or type 2 (W + Ww). From *leiten* (*to conduct*), we can derive *Leitfähigkeit, conductance* (the property of a material by virtue of which it allows current to flow through it when a potential difference is applied), and from *Leitung* (*conduction*) *Leitungsfähigkeit* (*conductivity*). The following terminations of this type may be listed:

-körper	(body, substance)	
	Festkörper	*solid*
	Sprengkörper	*explosive charge*
-gut	(goods, possession)	
	Mahlgut	*material to be ground*
	Schuttgut	*loose material*
-stoff	(material)	
	Baustoff	*building material*
	Brennstoff	*inflammable material, fuel*
	Farbstoff	*dye-stuff, pigment*
	Werkstoff	*raw material*
	Zellstoff	*cellulose*
-stück	(piece)	
	Mundstück	*mouthpiece* (trumpet)
	Werkstück	*worked article*
-werk	(product of some process)	
	Backwerk	*pastry*
	Pelzwerk	*furriery, furs*
	Schuhwerk	*footwear*
-zeug	(instrument)	
	Fahrzeug	*vehicle*
	Flugzeug	*aeroplane*
	Hebezeug	*lifting gear*
	Werkzeug	*tool*
-mittel	(means)	
	Bindemittel	*cement*
	Füllungsmittel	*additive*
	Lebensmittel	*food*

-glied	(member, link)	
	Bindeglied	*connecting link*
-gerät	(device, equipment)	
	Messgerät	*range finder*
-bau	(construction)	
	Maschinenbau	*mechanical engineering*
-teil	(part)	
	Bruchteil	*fraction*

Adjective-forming elements

Adjectival derivations include the following large groups: noun + adjective (WW); adjective + adjective (WW); and above all adjective + suffix (Ww). There is a high degree of correspondence between the N + A compounds in the two languages. The German propensity for fusing two words into one is matched in English by NA combinations. Examples are *feuerfest*, from *Feuer* (*fire*) and *fest* (*firm, enduring*), which can be replaced by *fireproof*; or the synonym *feuerbeständig*, from *Feuer* + *beständig* (*stable, steadfast, unchanging*); *wasserdicht*, from *Wasser* + *dicht* (*tight, impervious*) meaning *waterproof*, *watertight*; and *einwandfrei*, from *Einwand* (*objection*) + *frei* (*free*), with the English equivalent *unobjectionable*, *satisfactory*. The closeness of the parallels can be underlined by relating *feuerfest* to *fire-fast* and *einwandfrei* to *troublefree*.

Some of these combinations are analogous to the NN forms described above, in that the determinant at the end of the expression tends to become a suffix. Its independent nature as a word is no longer felt and it is used as a suffix. These constructions are very popular in technical writing:

-fähig	(capable of)	
	lebensfähig	*viable*
	leitfähig	*conductive*
-mässig	(mode, how something is arranged)	
	ordnungsmässig	*orderly*
	regelmässig	*regular*
	verhältnismässig	*proportional*
	zweckmässig	*appropriate*
-gemäss	(in accordance with)	
	ordnungsgemäss	*orderly, regular*

-reich	(rich in)	
	erfolgreich	*successful*
	sauerstoffreich	*rich in oxygen*
	wasserreich	*abounding in water*
-arm	(poor in)	
	CO_2-arm	*with low CO_2 content*
	wasserarm	*arid*
-los	(less)	
	erfolglos	*unsuccessful*
	wertlos	*worthless*
	wirkungslos	*ineffectual, futile*
-voll	(full)	
	bedeutungsvoll	*meaningful*
-artig	(like, of the type of)	
	gleichartig	*similar* (of the same kind)
	verschiedenartig	*varied* (of different kind)

All these derivations are of the NA and VA type. The AA combinations are used to express a more intense degree of the quality suggested by the main adjective, or a more specialized meaning. The word *grösstmöglich* corresponds to *greatest possible*; *altmodisch* to *old-fashioned*. These are of course wW forms.

Of greater significance is the large range of Ww adjectives. Among the principal morphemes used to derive adjectives in German are *-bar*, *-haft*, *-isch*, *-lich* and *-sam*. The termination *-bar* is used to transform verb stems into adjectives and usually has the sense of being able to do something. Its typical English equivalent is the suffix *-able*. Thus from *bewegen* (*to move*) (stem *beweg-*) we can derive *bewegbar* (*movable*), while *merken* (*to notice*) yields *merkbar* (*noticeable*). Less common are *-bar* derivations from nouns. One example is *fruchtbar* (*fruitful, fertile*) from *Frucht* (*fruit*). Adjectives can also be formed from other adjectives by the addition of *-bar*, as with *offenbar* (*evident*) from *offen* (*open*).

For part of the range *-lich* corresponds to *-like* and *-ly*. But English uses a number of other suffixes to convey the idea of likeness. Hence *bräunlich* is not rendered as **brown-like* but as *brownish*. The equivalent of *länglich* (**long-like*) is *elongated*; of *räumlich* (**space-like*) *spatial, spacious*; while *gründlich* (*basic, down to earth, fundamental*) means *thorough*. In general *-lich* conveys the notion of 'the property of'.

The sense of -*haft* is possession (having the property of), and -*ive* is sometimes the English equivalent. The adjective fehlerhaft is derived from the noun *Fehler* (*defect, error*) and corresponds to *defective, faulty* (*defect-having) while *vorteilhaft* from *Vorteil* (*advantage*) corresponds to *advantageous, favourable* (*advantage-having).

The -*bar* ending contrasts with -*lich*. The translation of *bewegbar* is *movable*, but in the sense of something that can be moved. That of *beweglich* is movable in the sense of flexibility, of something actually moving. *Förderbar* means *transportable* but *förderlich* means *conducive* (*to*). The sense of -*bar* is passive, of being acted upon by an external force, while that of -*lich* is rather a quality inherent in the object.

The suffix -*sam* expresses ability and tendency and is usually associated with human beings: *arbeitsam* (*industrious, active*); *betriebsam* (*industrious*).

The suffix -*isch* expresses origin and nature and is used in particular for derivations from proper names. It generally corresponds to English -*ish* (cf. *englisch, English*). Sometimes the English counterpart does not have the suffix, as in *afrikanisch, African*, but the meaning is easy to interpret. Foreign-language borrowings in German are rendered by the suffix -*ic* or -*ical* in English: *historisch, historical; physisch, physical; tragisch, tragic; statisch, static*.

A number of adjectival suffixes of foreign origin are also used in German: praktik*abel*, pol*ar*, regul*är*, kultur*ell*, hybr*id*, ration*al* (= according to reason), ration*ell* (= rationalizing industrial processes). These are easy to translate into English, except that one must guard against confusing similar forms such as *rational* and *rationell, formal* (relating to form) and *formell* (regarding social forms).

Adverb-forming suffixes in German include -*halb*, as in *ausserhalb* (*beyond, in addition to*); -*weise*, as in *beispielweise* (*by way of example*), contrasting with *beispielhaft* (*exemplary*); and -*wärts* as in *aufwärts* (*upwards*).

The above survey reveals that there is some degree of regularity in the process of word-building, though this is perhaps more true of German than of English. It is certainly not a haphazard process. There seems to be a constant drive to achieve logic and consistency in the structure of words, yet this is counteracted by other forces –

the forces of tradition and conservatism, by which old forms based on obsolete notions and classifications interfere with new attempts at order; and the forces of innovation, which introduce new terms into the vocabulary. These novelties enter the language in such a way that they upset the existing order. Language innovation is not controlled by linguists or logicians or engineers. This applies even in technical language, as we shall see, in spite of a closer control over vocabulary. It is clear that if we cannot achieve complete regularity in a single language, it will be even more difficult to obtain a regular correspondence between two languages.

CHAPTER 8

Sentence patterning

𝔊𝔊𝔊𝔊𝔊𝔊

THIS chapter deals with syntactic range in German and English. Its concern is with types of sentence, basic sentence patterns, rules of word order and types of clauses. The elementary fact is that a sentence is not merely the sum of the words that compose it. This can be demonstrated by something as simple as a combination of the words *he* and *is* (in German *er* and *ist*). We can produce from these:

> he is
> is he?
> *he* (and nobody else) is
> he *is* (He exists, he is not non-existent)

In speech we can increase the range of possibilities by means of different intonations and stresses. But this alone is enough to show that a sentence is more than the sum of its parts. It has order and structure. It has, moreover, a hierarchical structure. The words in a sentence stand in a definite relation to one another, and are in relations of subordination and superordination:

> 1   the man watched the dog
> 1a   the tall man watched the black dog
> 1b   the tall man watched the black dog attentively
> 1c   the tall man watched the black dog attentively for a long time

The two essential ideas in the above sentences are 'man' and 'watch'. We could strip everything else away and as long as we had 'man' and 'watch(es/ed)' we would still have an intelligible sentence. If we add 'the' to 'man' we make it clear that a particular member of the species 'man' is watching, not mankind in general. But 'the' is not essential; it is subordinate to man. In the same way, in 1a 'tall' is subordinate to 'man' and 'black' is subordinate to dog. In 1 'the dog' is subordinate to 'watched'. In 1b a further element is introduced: 'attentively'; this too is subordinate to 'watched'.

All language utterances display hierarchy, even a word when it consists of more than one morpheme. As we have seen, a word consisting of more than one morpheme is made up of a main element and a determinant element. The main element in 'watched' is 'watch', and '-ed' is a determinant signifying the past tense. With a lexical item such as 'noisiness' the structure is a little more complex. We have: a) *noise* + *y* [i] (a determinant that converts *noise* into an adjective, *noisy*); and b) a further determinant, *-ness* (which converts the adjective *noisy* into the noun *noisiness*). Thus *-ness* is the determinant of *noisy* and *-y* is the determinant of *noise*. The same principles of sequence and priority are found in all sentences.

The components of a sentence are words, but these are organized in groups according to hierarchical principles. These groups in turn are organized according to hierarchical principles within the sentence. A group of more than one word that functions as a word class (i.e. it can replace a word class in a sentence) is a phrase. A group of words organized in a relation of subject and predicate, requiring at least two word classes, N and V, is a sentence. A sentence that is part of another sentence is a clause. A clause may replace a word class. Thus we can have noun phrases (NP), verb phrases (VP), adjectival phrases (AP) and so on. Similarly we may have adjectival clauses, noun clauses and so on, but we can also classify clauses in other terms. Just as a word of more than one morpheme consists of main element and determinant, a phrase consists of a main word (head word) and modifying words. In 1a) 'the tall man' is a NP, with 'man' as head word and 'the tall' as modifiers. In 1c) 'watched attentively' is a NP, with 'watched' as head word and 'attentive' as modifier. We can also include 'for a long time' in 'watched for a long time', where 'for a long time' modifies 'watched'.

Although a sentence normally requires at least two elements – a noun and a verb – sentences with only one element are possible. Any type of word can be used, as in the exclamation: *Here!*, but a more usual type of one-word sentence is the command: *Come!* But the overwhelming majority of sentences in an utterance, and especially in a technical utterance, will have at least a subject and a predicate, and the subject will be a noun or noun phrase, the predicate a verb or verb phrase. The predicate may also contain an object complement (in 1) 'the dog').

The subject is always a noun or nominalized construction, such as an infinitive verb used as a noun. In German the subject is always in the nominative case. Both German and English possess several different types of sentence:

declarative or statement sentence (*The man watched the dog.*)
command or imperative sentence (*Watch the dog!*)
exclamatory sentence (*What a big dog that is!*)
interrogative sentence (*Is the man watching the dog?*)

The declarative sentence is by far the most common. But any sentence can be transformed without difficulty into another. We can do this even without changing the elements or their order, but simply by introducing the appropriate punctuation signals (or in speech, by intonation). The following examples (in German) are taken from the *Duden* grammar:

Sie kommen doch morgen (declarative: They are coming tomorrow)
Sie kommen doch morgen! (exclamatory: They are coming tomorrow!)
Sie kommen doch morgen? (interrogative: They are coming tomorrow?)

All sentences can be modelled on a relatively small number of patterns. Although there is no general agreement as to the exact number or nature of these, the divergences are not very great. The simplest is a NP-VP one (subject-predicate): *Die Sonne scheint* (*The sun shines*). The English sentence is bound by a rigorous word order in which the subject comes first and the predicate follows. Only in exceptional circumstances can this order be varied. The German word order is the same in the NP-VP sentence but it is not followed in a large number of more complex sentences. The English sentence can be thought of as typically linear, with the different classes (subject, predicate, direct object) following one another in a specific order of precedence that to an English speaker seems the logical order. It does in fact conform to the pattern of a logical proposition. But the sequence in German will frequently seem illogical, irregular and complicated to an English speaker. This difference is a major source of error in translation, and represents one of the major difficulties for the translator.

The most important sentence patterns can be listed as follows:

S-P (NP-VP)

S-P-O (NP-VP-NP)

  i) VP identificatory (*ist/is*): identifies the subject with the object

  ii) VP relational: the subject acts on the object

S-P-A (NP-VP-AP): identificatory; the subject is identified with the adjectival qualifier

S-P-AdvP (NP-VP-AdvP)

S-P-O-Oi (NP-VP-NP-Pp+NP)

S-P-PpO (NP-VP-PpO)

S-P-O-Adv (NP-VP-NP-AdvP)

This list can be extended if we include combinations such as S-P-O-PpO.

The S-P sentence is exemplified in *Die Sonne scheint*. The verb is intransitive; it cannot have a direct object. (In English, of course, *shine* can also be used transitively, as in *He shines his shoes*.) Other examples are: *der Hahn kräht* (*the cock crows*); *er singt* (*he sings*). Many impersonal forms beginning with *es* and of the type *es* + V (3rd person singular) belong to this category. In German verbs expressing weather conditions can normally be used only in this form: *es regnet, es schneit* (*it is raining/snowing*). This is a very common type of construction in technical prose: *Es wird in diesem korpuskularen Bild auch ohne Schwierigkeiten verständlich*. An English form related to this is of the type *there* + *be*. This can often translate the *es* + V sentence. For example the English version of the above sentence is: *With this corpuscular theory there is no difficulty in understanding that....* This form can also be used in German to express sensations, as in *es friert mich* (literally: *it freezes me), *es hungert mich* (*it hungers me). But these forms are now archaic, which is fortunate for translators, as there is nothing comparable in English. The more usual form nowadays would be *ich habe Hunger* (*I have hunger) = *I am hungry*.

Just as *there is* sentences can be replaced by a more direct S-P-O form, so *es* + V can also be translated in this way: *Es war damit das vollständige elektromagnetische Spektrum ... bekannt geworden* (*Thus the electromagnetic spectrum became known...*). Where the correspondence is closer, as with *Das Licht brennt*, the more suitable form in the TL will probably be one using a progressive tense rather than the simple present. Thus *the light burns* is probably better translated as *the light is burning.* Similarly *it is raining, it is snowing* are preferable to *it rains, it snows.*

The *es* + V phrase is a popular way of introducing a *dass* (*that*) clause or infinitive construction: *es ist nötig/möglich/wahrscheinlich/ sicher* (*it is necessary/possible/probable* [*likely*]/*certain*). It may also be possible to replace these in English by expressions such as *possibly, probably, it seems, it appears, evidently.*

The *relational* type of S-P-O sentence is an important pattern, though it is by no means the most common one. *Hans schreibt einen Brief = Hans writes a letter.* In German the direct object in a sentence like this must be in the accusative case (the article becomes *einen*). In English it must follow the verb.

The *identificatory* type of S-P-O sentence is more common and has a very high frequency in technical texts:

2a *Steroids sind organische Verbindungen*
2b *Steroids are organic compounds*

The verb in such sentences is usually *sein/to be.* It acts merely as a link between the two parts of the sentence. This is the simplest and most typical form of the identificatory sentence. Other sentences have a similar form, using for example the verb *sein/to be,* but in combination with, say, a noun that carries the notion of an action. Thus:

3a *Der Ausfall dieser spezifischen Wirkstoffe . . . ist meist Ursache für ernsthafte . . . Erkrankungen*
3b *Lack of these specific active constituents . . . is usually the cause of serious . . . diseases*

The action element in this sentence is really *ist Ursache, is the cause.* The actual verb is empty of lexical meaning and conveys only the meaning of 'verb'. It is structural, part of the scaffolding, and the relationship between S and P, especially that part of the predicate represented by *für ernsthafte Erkrankungen,* is signified by *Ursache, the cause.* This can be seen if we substitute a full verb and write:

3c *Der Ausfall dieser spezifischen Wirkstoffe bewirkt meist ernsthafte Erkrankungen*
3d *Lack of these specific active constituents usually causes serious diseases*

Thus although 3b) is a correct translation of 3a), 3d) is an alternative rendering that would also be acceptable in this particular instance.

Sentence 3 is obviously very different from sentence 2, which simply identifies (*Steroids are members of the class of organic compounds*). Another difference is that we can put sentence 3 into the passive without radically altering the meaning. In sentence 2 we cannot even reverse the position of the S and O: *Organic compounds are steroids* does not mean the same as *Steroids are organic compounds*. This is in fact untrue. We can sometimes reverse the order in German because of the case inflection. Thus *Einen brief schreibt Hans* is true if *Hans schreibt einen Brief* is true. *A letter writes Hans* is nonsensical in English, even if *Hans writes a letter* is accurate.

A sentence of this type can occasionally be reversed without affecting the meaning, however. *Karl ist mein Freund* is equivalent to *Mein Freund ist Karl* and to *Karl is my friend* and *My friend is Karl*. There is almost complete identification between the two parts of the sentence: *Karl = my friend*. This is an extremely important sentence type in technical texts and is perhaps expressed in its clearest form in mathematical equations. In grammatical terms, the relation between subject and object is one of apposition. A paraphrase of the sentence turned into a phrase would be *My friend Karl*. In this phrase *my friend* and *Karl* are of equal status; neither is subordinate to the other.

The identificatory sentence of the S-P-A type is exemplified in *Karl ist gut* (*Karl is good*). This does not mean that *Karl = good* but that *Karl* has the quality of being good. In the paraphrase, *the good Karl*, *Karl* is the head-word and *good* is subordinate. Here is an example of this in technical language:

4a *So sind . . . die unteren Gasschichten . . . sauerstoffreicher und CO₂-ärmer als die eigentlichen Verbrennungsgase*

4b *The lower gas layers . . . are richer in oxygen and poorer in CO₂ than the actual combustion gases.*

In the S-P-Adv sentence the adverbial element supplies information on the location of the subject, its duration, its cause and so on:

5a *Die Begriffe Atom und Molekül entstammen bekanntlich der Chemie*

5b *The concepts of atoms and molecules originated in chemistry*

5c *München liegt an der Isar*
5d *Munich lies on the Isar*
5e *Die Beratung dauerte zwei Stunden*
5f *The consultation lasted two hours*

The correspondences between these basic patterns in German and English have so far been fairly close. When we come to the S-P-O-Oi pattern there is greater divergence. The relationship between the verb and the predicate elements is more explicit in German than it is in English. The direct and indirect objects in English are indicated only by their sentence position, which can result in ambiguity. This may not be apparent in *He gave the boy the book*; but if we expand the sentence ambiguity results: *He gave the boy the book [which] he saw.* There is no obscurity in German since the relationships are indicated by the case endings. *Er gab dem Jungen das Buch* is perfectly clear as to the status of the different units: *dem Jungen* (dative, masculine, singular, hence 'to the boy'), *das Buch* (accusative, neuter, singular, hence 'the book', object of the action).

The dative object as exemplified in these examples is usually replaced by the indirect object in English sentences of this kind. The indirect object can be indicated by case endings in English only if it is a pronoun, as in *He gave him the book.* Ambiguity may be avoided by the use of 'to': *He gave the book to the boy*; the correspondence may be many-to-one and part of the meaning may be omitted in English. Thus *Der Sohn dankt dem Vater* (S-P-O [dative]) is translated as S-P-O: *The son thanks the father.* The dative object is treated as a direct object. But a dative object can exist in a German sentence as the only object, whereas an indirect object cannot exist as the only object in an English sentence. Thus we can say that the German dative object partly overlaps with the English direct object and partly with the indirect object. It is not fully equivalent to either. It corresponds to the indirect object. The sentence pattern S-P-O-Oi expressed in terms of word classes was given as NP-VP-NP-Pp+NP, corresponding to English usage. The true translation formula is rather: (SL) NP-VP-NP (acc) -NP (dat) = (TL) NP-VP-NP-Pp+NP.

These objects are governed directly by the verb and the verb type may determine the type of object. The verb *danken*, for example, must have a dative object. Other German verbs take only a prepositional object. The noun headword in the prepositional object is governed by the preposition. These restrictions do not apply in English, where the issue is whether to treat a construction as verbal (verb + preposition) and taking a direct object (as in *walk up* + *the hill* in *to walk up the hill*) or as verbal (verb only), taking a prepositional phrase as object (as in *walk* + *up the hill*. This difference may be significant, since certain German verbs and verbal forms (examples are those expressing aspect) correspond in English to prepositional phrasal verbs. One example is *erblicken*, which can in some contexts be rendered as to *catch sight of*. The equivalent of the expression *jemanden antreffen*, *to come across (some-one)* is a prepositional verb. The following sentence should probably be treated as containing a prepositional verb:

6a *The mill laboratory . . . made a tremendous impression on the millers of those days*

The verb can conveniently be paraphrased as 'impressed (deeply/tremendously)'. Thus 'made an impression' – 'impressed'. This is closer to the original German:

6b *Das Mühlenlaboratorium . . . beeindruckte die Mühlenfachleute damals tief*

The phrase forms a complete and independent lexical unit that can be used instead of the single verb. The words 'ran up' in *He ran up the hill* do not constitute such a unit. We can substitute, for example, the allied verb 'walk' for 'run': *He walked up the hill*, or the preposition (also associated in meaning – direction) 'down' for 'up': *He walked/ran up/down the hill*. This does not apply to *He ran up a bill*, where *ran up* is a single unit of meaning. *He walked/ran up/down a bill* does not make sense. Similarly, *He ran down his enemy*, but not *He walked up/down his enemy* or *He ran up his enemy*. 'Run up' and 'run down' in the latter examples are verbal units, but in the former they are verb + part of a prepositional unit ('up the hill', 'down the hill'). A guide to determining the nature of these phrases is whether the verbal unit can be replaced by a single verb *machen . . . leichter (erleichtern, facilitate)*; *give in (surrender)*; *give away* (*he gave his books away* but *he gave away the secret (betray, disclose)*).

In the S-P-O-Adv pattern the adverbial modification may on occasion be replaced as above by a single verb of the type that expresses a special mode of action (causation, initiation, intensification, reiteration and so on). The adverbial phrase may have a space, time or manner connotation.

COMMANDS

The command sentence in ordinary speech may carry a variety of emotional connotations. It may express a simple order, a demand, a request, a warning, an admonition, an invitation, a summons, a regulation, a decree, a challenge, a claim, a prohibition, a suggestion, a threat, or an instruction. In technical usage the emotional element is less common. The command form is mostly used in instructions, as in manuals and handbooks, in descriptions of experimental procedure, in legislative enactments. There may be emotional associations, as in safety warnings on apparatus and in instructions.

There are several types of command sentences. The most common is probably the V + Pp form. An example is: *Pass auf!* for which the English equivalent is: *Watch out!*, *Take care!* or *Beware!* The verb form in German is the second person singular, and may be expressed with the *-e* termination (*gehe!*) or without it (Geh!). This is also explicit in the English (*You watch out!*, *You take care!*, *You beware!*). It is encountered in footnotes such as the following:

7a *Siehe auch DIN* 18 031
7b *See also DIN* 18 031

Equally popular in German is the infinitive form: *Bandfach-Deckel öffnen, Open tape-chamber door.* The expression *bitte* may be included, as in invitations to subscribe to a publication:

8a *Bitte forden Sie das ausführliche Gesamtverzeichnis an*
8b (**Please request the extensive catalogue*)
   Write in for our extensive list
8c *Wenden Sie sich an Ihren Buchhändler oder schreiben Sie uns*
8d (**Turn you yourself to your bookseller or write to us*)
   Ask your bookseller or write to us

The transformation between SL and TL is grammatically Vi = V or *Bitte* + Vi + Pn (personal, second person polite form) = *Please* + V or *Bitte* + Vif+ Pn = V.

The following forms are also possible:

	D	E	
noun	*Vorsicht!*	*Watch out!*	V + Pp, V
	Achtung!	*Caution!*	N
adjective	*Schneller!*	*Faster!*	A
adverb	*Vorwärts!*	*Forward!*	Adv
past participle	*Aufgepasst!*	*Attention!*	N
question	*Kommst Du bald?*	*Are you coming (soon)?*	
	Pn + V + Adv	V + Pn + V (part) (Adv)	
exclamation	*Du gehst jetzt!*	*You're going now!*	
	Pn + V + Adv	Pn + V + Adv	
passive impersonal	*Jetzt wird geschlafen!*	*Now go to sleep!*	
	Adv + *sein* + V (past part)	(Adv + V + Pp + N) Adv + VP	
infinitive with *zu* + *haben* or *zu* + *sein*	*Die Tür ist sofort zu schliessen!*	*The door is to be closed instantly!*	
	NP + *sein* + Adv + *zu* + Vi	NP + *be* + *to* + *be* + V (past part) + Adv	

It will be noticed that there is a greater variety of imperative forms in German than in English. In general the types most often used in German are the second person verb, the noun, the infinitive and the infinitive with *zu* + *haben/sein*. The English reformulation is usually verb (sometimes preceded or followed by *please*), noun and must + participle (passive). The correspondence is not necessarily one-to-one. With the second person verb it is usually one-to-one, but infinitive (German) can also be rendered by a verb (English). A German noun may be rendered by an English noun or by a verbal phrase. The infinitive with *zu* may be in a one-to-one correspondence but it can also be rendered by *should*, *must* forms. The German *sein* + *zu* + infinitive cannot have a one-to-one correspondence in English, where the form must be passive (*be* + *to* + *be* + V [part]).

English has a polite form of imperative that does not exist in German. We have a one-to-two correspondence in the following prohibition:

9a *Prägeband nicht nach vorn ziehen*
9b *Be sure not to pull tape forward by hand*

The German does not differ from the sentence 7a type, but we cannot write in English:

9c *Tape not to pull forward

We could use the alternative:

9d Do not pull tape forward

One of the functions of *do* for which there is no German counter-part is to contribute to the formation of negative commands (prohibition) in this way: *Pull tape forward*: *Do not pull tape forward* (*Pull not tape forward*).

EXCLAMATIONS

The exclamatory sentence, which is nothing if not emotive, is rare in technical texts, but it is of some importance as an eye-catcher in advertising literature. In its simplest form it may consist of a single word, like *Feuer! Fire!* A typical sentence would be: *Wie schnell ist er gelaufen! How fast he ran!*

QUESTION SENTENCES

The interrogative sentence is distinguishable from a declarative sentence by its word order, and sometimes by the use of special interrogative words. It does not appear often in scientific and technical literature, which prefers affirmative statements, but it is used in questionnaires and in check lists.

A basic declarative sentence would have the pattern S-P or S-P-O. A basic question sentence would reverse this order to P-S or P-S-O. A sentence can, however, be converted into a question merely by placing a question mark at the end of it. Such sentences generally have emotional overtones and are uncommon in the type of texts that are our main concern. They are also uncommon in English. The following is the more usual:

10a *Sind alle Arbeitsplätze markiert?*
10b *Are all work places clearly marked?*

This follows the P-S pattern, and we note that in both SL and TL sentences there is a separation between the auxiliary verb and the participle, which comes at the end of the sentence. A difference occurs in the following examples:

11a *Erfolgt regelmässig Ausbildung der Lehrlinge durch einen geeigneten Meister?*

11b *Do the apprentices receive regular training from a qualified master?*

11c **Follows regular training of the apprentices by a qualified master?*

We immediately recognize 11c as an un-English word order. Inversion requires the auxiliary *do* at the beginning of the sentence. This auxiliary is used only for negation (see 9d) or inversion or emphasis. The German sentence form for questions has the formula V+NP or, as in 11a, V+NP+PpP, whereas the English counterpart is Do + NP + V + NP + PpP.

A question sentence may be introduced by an interrogative pronoun: *wer* (masc and fem) corresponding to *who*; *was* (neut) equivalent to *what* or *which*; *welcher*, *which*. The question word comes at the beginning of the sentence, Pn + V + NP (S-P-O). Where a verb phrase consisting of an auxiliary together with an infinitive or participle is used, the latter comes at the end of the sentence, Pn + Vaux + NP + (NP) + Vi/part. The English equivalent of this is Pn + Vaux + NP + Vi/part + (NP).

12a *Wem hat Hans gestern in der Vorlesung das Buch gegeben?*

12b *To whom did Hans give the book at the lecture yesterday?*

A somewhat more complicated question sentence following the same essential pattern is given below:

13a *Was bedeutet es nun, wenn alle Zustände negativer Energie bis auf einen besetzt und alle Zustände positiver Energie leer sind?*

13b *What is to be understood if all states of positive energy are vacant and all states of negative energy but one are occupied?*

The form of 13a is Pn + V + N(PN) + Adv + *wenn*-clause and that of 13b is Pn + VP + *if*-clause. This will be better understood in terms of complex sentence structure.

COMPLEX SENTENCES

Complex sentences consist of more than one clause. The two or more clauses may be independent, and the relation is then one of co-ordination. Usually they are associated by means of a linking word such as *und* (*and*), *oder* (*or*), *entweder . . . oder* (*either . . . or*), *aber* (*but*) and similar expressions. The correspondences between the two languages are very close and problems rarely arise here. These sentences are anyway not very likely to occur; subordinate clauses are far more common, particularly in technical language, where conditionality, cause and effect, result and so on are so much part of the thinking. The relation may be one of simultaneity (*während, while*), causality (*weil, because*), consequence (*dass, that*), conditionality (*wenn, if*), concessiveness (*obwohl, although*), instrumentality (*dadurch, through which*). In addition we have a host of relative clauses or attributive clauses (which are probably the most numerous) and clauses that operate like word classes (subject clause, object clause and so on).

The word order of a subordinate clause in English almost invariably differs from German. In German, to begin with, the verb comes at the end. This is not so in English. A German word order will be S-P-Comp+Cn+S-Comp-P. The English equivalent is S-P-Comp+Cn+S-P-Comp.

14a *Die Technologie der Herstellung von autoklavbehandelten Erzeugnissen vermisst bisher . . . notwendige Unterlagen der Grundforschung†, da manche Voraussetzungen . . . ausschliesslich auf empirischen Kenntnissen beruhen*

14b *The technique of processing autoclaved products has hitherto lacked certain essential basic research data since some assumptions . . . are based entirely on empirical knowledge*

The skeleton of 14a is *Die Technologie . . . vermisst . . . Unterlagen da manche Vorasussetzungen . . . auf empirisch Kenntnissen beruhen* (S-P-O-Cn-S-PpO-V). The corresponding framework of 14b is *The technique . . . has lacked . . . data since some assumptions . . . are based . . . on empirical knowledge* (S-P-O-Cn-S P-PpO).

A further example again illustrates this:

15a *Die Lösung dieser Aufgabe würde ja die Lösung des Vielteilchenproblems der klassischen Mechanik voraussetzen, weil die Elektronen alle aufeinander mit Coulomb-Kräften wirken.*

† *Grundforschung* is, of course, incorrect and should be *Grundlagenforschung*. These examples are excerpts from actual texts in which some of the language leaves something to be desired. This is one of the 'miseries' of the technical translator – that he very often has to cope with misprints, omissions and ungrammatical utterances in addition to the normal occupational risks. The original sentence would have read better as follows: *In der Technologie der Herstellung von autoklavbehandelten Erzeugnissen vermisst man bisher . . . notwendige Unterlagen der Grundlagenforschung, da manche Voraussetzungen . . .*

15b *A solution of this problem would presuppose a solution of the many-particle problem of classical mechanics, because the electrons all act upon one another with Coulomb forces.*

The skeleton of 15a is: *Die Lösung* (S) . . . *würde* (P) . . . *die Lösung* (O) . . . *voraussetzen* (P), *weil* (Cn) *die Elektronen* (S) . . . *aufeinander* (O) . . . *wirken* (P). That of 15b is: *A solution* (S) . . . *would presuppose* (P) . . . *a solution* (O) . . . *because* (Cn) *the electrons* (S) . . . *act* (P) *upon one another* (PpO). We find the same divergence in the subordinate clause structure of Cn-S-O-P-: Cn-S-P-O.

This general rule applies also to relative clauses (*der, die, das, which* clauses). The German pattern is S-P-Comp-*der*-Comp-P and the English S-P-Comp-*which*-P-Comp. There are, however, alternative translations:

16a *Die Gesetzmässigkeiten der Röntgenspektren sind es ja gewesen, die das obige Bild vom Atom am deutlichsten nahegelegt haben*

16b *The laws governing the X-ray spectra gave the clearest indications for the described picture of the atom*

16c *The laws of the X-ray spectra were those which most clearly suggested the above picture of the atom*

Sentences 16b and 16c are alternative versions of 16a, and 16c obeys the general rule. We might say that 16b is a stylistic variation. (For further discussion on this point see Chapter 14.) However it is necessary to note that a subordinate clause of this type can be replaced by another structure, in which the verb of the clause is incorporated into the verb phrase of the main sentence (*gave . . . indications* instead of the functional verb *sein/be* of the original and incorporating the concept of the lexical verb *suggest/ nahelegen*). The English version of a relative (*der, die, das*) clause is very often an adjectival phrase:

17a *Es soll hier lediglich . . . auf die Möglichkeiten, die sich für die Strukturaufklärung . . . bieten, hingewiesen werden*

17b *It is only intended . . . to point out the possibilities provided . . . in structural investigation*

In 17b a participle replaces the Cn+VP construction (*which are provided*). The basic rule, however, remains that *der, die, das* clauses can be rendered in English by *which* (*that*) clauses in which the main verb follows the pronoun.

Clauses that can function in the place of a subject, object, and so on, are an important group, although they are less frequent than relative clauses. Of these the object clause is the more common. It has the form S-P-O (*dass*+S-O-P), with the English equivalent S-P-O (*that*+S-P-O).

18a *Das internationale Entwicklungstempo . . . fordert, dass die . . . Fragen sehr schnell beantwortet werden müssen*
18b *The rate of development internationally . . . requires that the question . . . should be answered without delay*

Again, a phrase may replace the clause:

18c *The rate of development internationally requires the questions . . . to be answered without delay*
18d *The rate of development internationally requires a swift answer to the questions*

The subject clause relationship is much the same in both languages. The pattern is *dass*+S-O-P+P-O for German and *that*+S-P-O+P-O for English:

19a *Dass die Entscheidung auch heute noch nicht immer zugunsten des Zentrifugaldekanters ausfällt, liegt an dem bequemeren und universelleren Anwendbarkeit der Schälschleuder (Dass die Entscheidung . . . nicht . . . zugunsten des Zentrifugaldekanters ausfällt, liegt an der . . . Anwendbarkeit . . .)*
19b *That nowadays the centrifugal decanter is not always decided upon is due to the more convenient and more universal applicability of the semi-continuous solid bowl centrifuge* (That . . . the centrifugal decanter is not . . . decided upon is due to the . . . applicability . . .)
19c *If nowadays the centrifugal decanter is not always the choice this is due to . . .*

The remarks about the object clause are applicable here too. It should also be noted that the conjunction need not be *dass/that*, but can be *ob, whether/if; wer, who; was, which.*

Conditional clauses merit special attention because, like causal clauses, they are particularly important in technical literature. A large number of scientific statements are of the *wenn . . . dann, if . . . then* type, although these words may not actually be used. In German the pattern is *wenn*+S-(A)-P+*dann*+P-S. In English, it is *if*+S-P-A,*then*+S-P.

20a *Wenn der Rechercheur† seine Neuheitsvergleiche bei geringstem Zeitaufwand ausführen will, hat er zu berücksichtigen, dass die internationale Patentklassifikation aus Unterlassen bzw. Fachgebieten besteht . . .*

20b *If the searcher wishes to carry out his novelty comparison in as short a time as possible, he must remember that the IPC consists of sub-classes or disciplines . . .*

The framework of 20a is *wenn* + *der Rechercheur* (S) *seine Neuheitsvergleiche* (O) *ausführen* will (P) + *hat* (P) *er* (S) . . .; *and* of 20b *if* + *the searcher* (S) *wishes to carry out* (P) *his novelty comparison* (O) +*he* (S) *must remember* (P) . . .

Conditional statements are also formed in German when the main clause starts with the verb instead of *wenn*: P-S-O+P-S-O(A). English cannot do this:

21a *Sind Durchfahrten zugleich Verkehrswege, soll ein Sachverständiger hinzugezogen werden, Anforderungen ggf. höher*

21b *If passages are also traffic routes, an expert should be consulted: requirements may be higher*

But one type of English sentence expressing a condition does resemble the German in starting with a verb:

22a *Rührte das Moment von der Bahnbewegung des Valenz-elektrons her, so könnte man die Quantenmechanik der Bewegung des Atoms im Magnetfeld ohne Schwierigkeiten entwickeln*

22b *Were the moment to originate in the orbital motion of the valency electron, there would be no difficulty in developing a quantum mechanics for the motion of the atom in the magnetic field*

This sentence could also be rendered by 'If the moment originated . . .'

Related to the conditional sentences are those that begin 'Suppose that', 'Assume that' and so on. One form in German is an utterance beginning with *Es sei . . .*

23a *Es sei die Strahlung in einem kubischen Hohlraum vom Volumen $V = a^3$ mit metallischen Wanden eingeschlossen*

23b *Suppose the radiation to be enclosed in a cubic cavity with the volume $V = a^3$, with ideally conducting metallic walls*

† *Rechercheur* is a term used in documentation and should not be confused with 'rearcher' or 'research worker' as used in science.

The 'suppose' sentence may be identical with the conditional, as with *Nimmt man nun nach Einstein an, dass . . .*: *If we assume with Einstein that . . .*

NEGATIONS

Any affirmative clause can be negated in German by the addition of *nicht*. *Ein (harmonischer) Oszillator der Frequenz v kann nicht jede beliebige Energie besitzen . . .* The English version of this statement (the beginning of the Planck quantum hypothesis) is: *A (harmonic) oscillator of the frequency v cannot possess an optional amount of energy. . . . Kann + nicht* corresponds closely with the fused *cannot*. In German a group of words can intervene between *kann* and *nicht*, but this cannot happen in English:

24a *Energie kann also von diesen System nicht kontinuierlich aufgenommen oder abgegeben werden, sondern . . .*

24b *The system cannot take up or give up energy continuously, but . . .*

24c *Energy cannot be taken up or given up from this system continuously, but . . .*

24d **Energy can from this system not be taken up or given up continuously, but . . .*

Thanks to its position at the end of the sentence, the negating element negates the whole sentence: *ich komme, I am coming*; *ich komme nicht, I am not coming (*I come not)*. When the negator is the subject of the sentence, as in *niemand kam*, the pattern is the same in English, *nobody came*. But when it is the object and the predicate consists of auxiliary and infinitive or participle, the word order does not correspond: *ich habe nichts gefunden*: *I found nothing* or *I did not find anything (*I have nothing found)*. The use of *do* for negation in English has no counterpart in German.

EXTENDED PHRASES

One of the bugbears of translating from German is its predilection for extended adjectival constructions that are not possible in English. The phrase *Der der Theorie des Compton-Effektes zugrunde liegende einfache Stossmechanismus* is perfectly normal in German. Literally it means: **The the theory of the Compton Effect underlying simple impact mechanism*. Once we understand the nature of the construction it need present no terrors. First, we see that the head-word is *Stossmechanismus* and that *der einfache Stossmechanismus* forms a unit (Art + A + N); using our rule for transforming German fused compounds (NN —N + N), we render this in English as *the simple impact mechanism*. Working backwards we invert *zugrunde liegende*, and say *lying at the base*. The remainder, *der Theorie des Compton-Effektes*, also forms a unit, *the theory of the Compton Effect*. Thus the simple impact mechanism underlies the theory of the Compton Effect. The translation of the phrase might then be *The simple impact mechanism on which the theory of the Compton Effect is based*.

This provides a model, and a rule can be inferred for translating extended participial adjectival phrases from German into English – by means of a relative clause. Sometimes, however, this is not necessary and one adjectival phrase may be translated by another, though a change in word order will inevitably be necessary. In the phrase above the structure is as follows: Art′+Art″+ NP″+VP(Adv+Vgerund) + NP. In English it is rendered by a phrase: *Mit den aus Eichmischungen erhaltenen Werten für Extinktion und Konzentration* Pp+Art′+Pp′+NP (Vpart+N) + PpP″. If we translate this into English, group Pp + Art′ + NP′ (inverting order of words in phrase to N+Vpart), we have **Mit den Werten erhalten*, followed by PpP′ and PpP″ in that order, hence **Mit den Werten erhalten aus Eichmischungen für Extinktion und Konzentration*. This is now an English word order, which can be rendered as: *With the values obtained from standard mixtures for extinction and concentration*. The actual translation was: *With the values for extinction and concentration obtained from standard mixtures*, which is better and clearer, but not obligatory.

The main point is to overcome the divergence in word order. The sensitive points are the position of the article in relation to its noun and the verbal element that must be inverted in the phrase. The article in English tends to be as close to the noun as possible. It is theoretically possible for an infinite number of adjectives to be inserted between article and noun on the lines of 'the big, dark, tall, elderly . . . man' but in practice this is rare, even for technical prose. (But note that technical prose is more Germanic in this respect than ordinary speech and can tolerate tapewords like *Pulse Time Constant of Fall*.)

CHAPTER 9

Lexical structure

𝕊𝕊𝕊𝕊𝕊𝕊

THE division between grammar and vocabulary is seldom absolute. The vocabulary of a language may be considered to be a collection of units that are used in accordance with a series of rules. These rules are known as the grammar of the language. We cannot think of the vocabulary of a living language without a grammar accompanying it, or of a grammar without a vocabulary. Each presupposes the other. Even artificial languages and codes have a grammar and a vocabulary.

Grammar can perhaps be defined as follows: grammar seeks to account for the greatest number of language facts by means of the smallest number of rules. No set of grammatical rules fully accounts for everything. Similarly, vocabulary is not merely an inert list but also displays structural features. In general the term 'grammar' as used here refers to the rules of formation, transformation and combination of the lexical units – the units comprising the vocabulary of a language. But not all the combinations, formations and transformations can be dealt with in terms of a limited set of grammatical rules. A further consideration, and one of the utmost importance, is that a living language is not an artefact. It undergoes a continual process of growth and change. Its grammar is changing, albeit slowly, and, more perceptibly, its vocabulary is constantly being transformed. This fact creates the greatest barrier to a science of translation. It also makes it necessary to understand grammar and vocabulary in two different senses: the grammar recorded in books and the grammar actually used by the speakers of the language. The same is true of vocabulary. There is a relation between the two of course. The average user of language seeks to obey what he believes to be 'the rules of usage' (the recorded grammar and vocabulary) but in fact the textbooks always and inevitably lag behind the language used for communication in practice. 'Rules of usage' are important since, without general agreement, communication would not be possible, but we must bear in mind that these rules do not conform completely to practice. Much technical writing in particular is 'bad German' or 'bad English' from the point of view of the rules, and indeed from other points of view.

The main difficulty underlying any attempt to discover system in a language's vocabulary is that there does not seem to be any. In contrast to the situation with grammar, there seems to be a multiplicity of rules, each accounting for a few facts, indeed, it hardly seems worthwhile to call them rules at all. This is certainly true if we consult a dictionary in the hope that it will do the same kind of work for us as a grammar book does. But a dictionary is not the same as a vocabulary, though it does of course contain a certain amount of vocabulary, depending on its size and type. But the organization of items in a dictionary hardly corresponds to the organization of an individual's vocabulary in his own mind. The words in a dictionary are arranged in alphabetical order. This is convenient, but it is not natural or necessary or even logical. It is merely the method we are used to.

If we pursue grammatical analysis beyond a certain point, we begin to approach the region of vocabulary organization. The contrast between categories such as 'animate' and 'inanimate'; the notion of 'aspect' in verbs; morphological contrasts – all these exist on the borderline between grammar and vocabulary. At the extremes we can distinguish fairly easily between the two approaches. The functional verbs are treated as grammatical without any confusion, and some lexical phenomena are clearly best treated as such. But the two also flow into one another. Rigid ideas about the separation of the grammatical and the lexical are a great handicap in translation, where transpositions between grammatical categories have to be performed and grammatical forms must also be replaced by lexical items. These are two ways of accounting for the data. Sometimes the one approach is the better, sometimes the other.

The unit of vocabulary, as of grammar, is the word. The morpheme is a smaller unit and the word group or phrase may constitute a larger unit, but the word is undoubtedly the most important here. The others can be viewed as secondary units. It is difficult to define a word and my approach will be a pragmatic one, accepting the common belief that there are such entities as words and that these are easily recognizable and identifiable. Since we are concerned with written language, a great deal of the difficulty in giving a precise definition of the word disappears. The written or printed word is easier to identify than the spoken one, and in literate societies part of education is teaching members of the community to recognize and identify words. The word is not, however, identical with the lexical unit, which may be a word, a morpheme or a group of words. One of the difficulties in defining a word arises from the transformations it may undergo. Are *boy* and *boys* the same word, different words or different forms of the same word? What of *man* and *men*, *sing* and *sung*? The term 'lexeme' is used to express the relationship between *man* and *men*. *Man* and *men* are different words but variants of the same lexeme.

The word is the lexical item most familiar to us but another lexical item is the morpheme. An example is *zer-* (+ V), which expresses, as we have seen, separation, destruction and disintegration. A phrase consisting of more than one word, like *resistance-capacitance-coupled-amplifier* or *guinea pig*, can also be a lexical item.

A word has a central meaning and related meanings. *Electrify*, for instance, is a verb with two groups of meanings: the first meaning is equip with or supply electricity, the second is an extended meaning, a derived or metaphorical meaning, and designates the causation of a human emotion: 'to excite intensely or suddenly as if by electric shock'.[1] In its central meaning the verb can be used only with inanimate objects (*to electrify a town*), but in the related meaning it is used only with animate creatures (*they were electrified by the news*). The verb *elektrifizieren* in German can be used only with inanimate objects. It is therefore in one-to-two correspondence with *electrify*. There is no problem in translating from German to English. It simply means that the English equivalent *electrify* must be used with an inanimate object. German here shows greater precision than English in its vocabulary.

The acquisition of extended meanings is one of the features of language growth and change, but technical language aims to reduce and obviate this extension and to eliminate related meanings already in existence wherever possible. But then of course ordinary language, in which these processes occur more or less unchecked, plays a considerable part in technical texts.

The extent to which related meanings may proliferate can be gauged from an example like the simple word *table*. Its central meaning in English is an article of furniture. Its related meanings include *tableland* (S *mesa*), *plateau, company, food, list* and (as a verb) *to present* (*a report*).

A lexical item does not correspond directly to a thing. It may denote an object in the external world, such as a table, or an imaginary thing, such as a centaur, or an abstraction, such as history. It also designates. Designation relates to what the majority of the users of the language believe the word means. The designation of *Walfisch* is (or rather was) a sort of fish, (though probably few literate people today believe this), as against the scientific denotation (mammal, member of Cetacea). Designation is consensual, based on the agreement of the speech community, for at the time when the term *Walfisch* was coined, all creatures that swam in the sea were considered to be fishes.

Since each speech community has its own agreements on designations, and since, in addition to its central meaning, a word acquires a number of related meanings, it is not surprising that the related meanings of words in different languages tend to diverge, though they may correspond in the central meaning.

In the realm of vocabulary, one-to-one correspondence is rare. Words overlap in meaning, but no word in one language agrees in its range with that of another. The range of *elektrifizieren* is, as we have seen, in one-to-one correspondence with *electrify*, although both verbs are constructed in the same way: a Latin root (originally derived from Greek) *electr-* and *-fizieren* (from Latin *facere* = *to make*), *-fy* (with the same ultimate origin); but *electrify* has acquired a number of related meanings that the German word has not. Similarly, German has a distinctive verb for particular types of grinding: *schleifen* (to grind knives for sharpening); *mahlen* (to grind flour); *knirschen* (to grind one's teeth). English has the one word *grind* for all these meanings. Conversely, *schleifen* can have the following meanings: *grind* (knife); *raze* (fortress); *cut* (diamond); *drag* (sack); *trail* (dress).

Attempts are still being made to establish systems in vocabulary. These include the idea of setting up minimal units of vocabulary, which would be more fundamental than the word and would play the role in language that atoms (or even sub-atomic phenomena) do in the physical sciences. In practice these have tended to amount to nothing more than establishing the meanings of morphemes, especially bound morphemes like *zer-*.

One of the most obvious types of lexical organization is found in oppositions like *schwarz* (*black*)/*weiss* (*white*); *gross* (*great*)/*klein* (*small*); *Vater* (*father*)/*Mutter* (*mother*); *Vater* (*father*)/*Sohn* (*son*). Affirmation/negation is another type of opposition and is found in morphological elements such as the *un-* prefixes in German and English (*annehmbar*/*unannehmbar*: *acceptable*/*unacceptable*). But the prefix *un-* does not correspond one-to-one like this throughout. Apart from the existence of similar negating prefixes like *im-* in English (*wahrscheinlich*/*unwahrscheinlich*: *probably*/*improbably*) and *in-* (*unbeugsam*: *inflexible*), the corresponding form in the TL may be a positive one. In some contexts *unübersehbar* corresponds to *illimitable* or *boundless*, but in others its equivalent is *vast*. Similarly, *unverblümt* is likely to be replaced by *blunt* or *plain*. The choice is determined by the surrounding context or situation, but a guiding principle is that there is often a positive form, close in meaning to the negative form, in English, as with *illimitable* and *vast*, *inflexible* and *rigid*, *uncamouflaged* and *plain*.

Among the simplest forms of opposition are those based on gender or sex, as F *sec* (masc), *sèche* (fem), *dry*, which has no counterpart in English, since English adjectives do not inflect for gender. However certain adjectives are used only with a particular sex. We say *a pretty girl* but hardly *a pretty man*; the neutral *a pretty state of affairs* is normally allowed, yet *a pretty ship* is unlikely, although *ship* is viewed as feminine in English. Certain adjectives, like *handsome*, are not outrightly masculine and can be used with females; but they do still have a definite male tinge.

An instructive example is *bull*/*cow*. Two ideas are contained in *bull* – the idea of the species (cattle = bulls, cows, calves) and maleness. Similarly cow = cattle + female. In German this is likely to be expressed in a Ww form:

Engländer	English + ø (i.e. unmarked = male)
Engländerin	English + female

In English the structure will be WW:

Englishman	English + male
Englishwoman	English + female

The first (male) form in German is the unmarked one and the second is marked for femaleness by the suffix -*in*. Sometimes the English expression may also be on the Ww pattern like the German:

| *Löwe* | lion species + male | *lion* |
| *Löwin* | lion species + female | *lioness* |

But there is not the same regularity in English as in German.

A more abstract opposition is that exemplified by the German modal auxiliary verbs *können* and *dürfen*. The following features constitute *können*: the idea of possibility; the idea that this possibility is material; and the suffix *-en* (which in conjunction with the fact that the initial letter *k* is in lower case indicates that it is a verb, infinitive form). The following pattern can be shown:

| *können* | possibility × material × verb infinitive |
| *dürfen* | possibility × moral × verb infinitive |

The elimination of the common suffix *-en* leaves the contrast between the roots **könn-* and **dürf-*:

| *könn-* | possibility + material |
| *dürf-* | possibility + moral[2] |

This corresponds roughly to the contrast in English between *can* and *must*, but it does not follow that these can always function as translations of the German verbs.

The same analysis can be applied to examples of central and related meaning like *elektrifizieren* and *electrify*:

elektrifizieren (C)† electric + make + inanimate
electrify electric + make + (C) inanimate (D)† electric + make + animate

The following French counterparts help to clarify this.

électrifier	(C)	electric + make + inanimate	*electrify*
électriser	(C)	electric + make + inanimate	*electrify*
électriser	(D)	electric + make + animate	*electrify*

The German is in a relation of one-to-two correspondence to English, but the French is a two-to-one correspondence (strictly speaking three-to-two).

A word may be considered as a collection of features. In a particular situation and in the context of other words, some of these features are singled out. *Table* is defined (*COD*) as 'Article of furniture consisting of flat top of wood or marble etc and one or more usually vertical supports especially one on which meals are

† C = central meaning, D = derived meaning.

laid out, articles of use or ornament kept, work done or games played...'. This is the central meaning of *table*, but it is made up of:

i	article of furniture	function
ii	flat top	shape
iii	vertical support(s)	structure
iv	for meals, articles of use, ornaments, work or games	function

It therefore contains at least three features, any of which may be emphasized. In the term *breakfast table* the function is emphasized; a breakfast table might be a packing case covered by a cloth, or a desk. Similarly the general function i or the specific function iv may be emphasized. Where iv is emphasized the resemblance to other tables may be tenuous, as between *dressing table* and *ping-pong table*. Thus:

	i	ii	iii	iv
dressing table	+	+	+—*	+
ping-pong table	—	+	+	+

(*A dressing table may be built into a wall and may not have legs.)

Some of the derived meanings listed in the *COD* are:

2 part of a machine-tool on which work is put to be operated on
3 slab of wood, stone, etc
4 matter written on this (Ten Commandments, Twelve Tables)
5 level area, plateau
7 flat surface of gem, cut gem with two flat faces
8 palm of hand, especially part indicating character or fortune
10 company seated at table
11 list of facts, numbers, etc., systematically arranged esp. in columns, matter contained in this, etc.

It is clear from this that not all these items have common features, but of the nine only seven share the flat top. The company seated at the table is a derived meaning, and 11 might be regarded as

	i	ii	iii	iv
breakfast table	+	+	+	+
ping-pong table	−	+	+	+
2	−	+		+
3	−	+	−	
4	−	+	−	−
5	−	+	−	−
7	−	+	−	−
8	−	+	−	−
10	−	−	−	−
11	−	−	−	−

something flat. An examination of the possible meanings in German *Tisch* makes it clear that the range will diverge somewhat, though chiefly in the kind of combinations that *Tisch* can enter into. The following can be excluded: 2, 3, 5, 8, 11. German has another word covering part of the range, *Tafel*. This would correspond to 3, 4, 11. A third word, *Tabelle*, corresponds to 11 when it is used for mathematical tables.

It is therefore often futile to ask for the meaning of a single word in isolation, since this will depend on which of the various possible meanings one wishes to select. It is true that here again German is rather more specific than English. A three-to-one correspondence narrows down the possibilities, but there is still room for doubt. The context is a necessary basis for choice, plus in some cases the set of circumstances in which the context occurs.

He sat down at the table involves the central meaning, if we allow the various possibilities within that meaning range (breakfast table, dressing table and so on). *The hotel offers a good table*, however, presents a derived meaning (i –; ii –; iii –; iv –). The expression *Astronomie, Tafel 1* suggests another (11). Similarly a reference to *die folgende Tabelle*, followed by a diagram giving figures for world production of a commodity, implies the same selection (11): *the table below*. Within a technical text this comes close to being a central meaning for *table*. Meaning shifts in this way in the course of time, so that a central meaning becomes the related one and vice versa. The original meaning of *crane* was a bird, and the associated meaning of lifting equipment was derived from this. Today the expression *crane* is as likely to evoke the idea of the equipment as that of the bird, since in most situations the equipment is likely to be the most relevant. An analogous word is *nut*. It is difficult to say which meaning is central nowadays. There is no such problem in German: *Nuss* is the edible nut; *Schraubenmutter* is the nut associated with a bolt. The correspondence is two-to-one. Again, no translation difficulty arises in English, though one must watch out for ambiguities. Note the utterance: *What caused the damage? A couple of nuts*. This has three possible interpretations!

Lexical range can be a very subtle matter and it is not easy to lay down rules, since there are no clear-cut demarcations. Words overlap and diverge in a way that may seem to be quite arbitrary. *Er sitzt am Tisch* translates easily enough as *He sits at the table*; *Dieses Hotel bietet einen guten Tisch* translates equally well as *The hotel offers a good table*, since the metaphor and derived meaning correspond. But the German expression may very well be *Dieses Hotel bietet gute gepflegte Küche/bürgerliche Küche* (*This hotel offers good well-tended/middle class/civic kitchen/fare*). The selections are made from an entirely different lexical area.

Specializations in meaning may be observed in many of these examples. German uses *essen* (to eat) only for human beings and *fressen* for animals. English usage contrasts with this: *eat* (general expression) and *devour* for eating like a beast or ravenously. *Devour* corresponds to a certain extent with *fressen*, but not entirely. Similarly the English translation for *Scheibe* (disc) will vary according to the thickness of the object or its function.

Scheibe	flat	*disc*
	three dimensions	*slice (of bread)*
	mechanical function	*sheave, pulley*[3]

Specialization may arise when the language does not provide the necessary lexical items by compounding, as in *Fensterscheibe* (pane). The general word *Zapfen* suggests something cylindrical or conical. There are many translations for this in English, depending on material or function.

Zapfen	(cone, stopper)	*cork, tap, plug, pin, peg, etc.*
Drehzapfen	(turn+)	*pivot*
Eiszapfen	(ice+)	*icicle*
Tannenzapfen	(fir+)	*fir cone*

A feature of great importance that emerges from some of the above examples is that of hierarchical classification in the vocabulary. *Hyponymy* implies the inclusion of a specific term within a more general term. *Flower*, for instance, is the more general or generic term in a group to which *blossom* and *bloom* belong. The class of flowers can be organized in different ways. It can be so arranged that it consists of daffodils, daisies, roses and so on. It can also become itself a member of the class of plants. *Meat* is a member of the class of *flesh*, but it is also a member of another class, that of *food*. These groupings bear a resemblance to the terminological systems of technical vocabularies but the latter tend to be more logically ordered. They have in fact been deliberately created on logical lines, whereas the hierarchies in the general vocabulary have arisen in a more haphazard manner.

In ordinary language *animal* is usually taken to be identical with *mammal*. The average person does not think of a bird or fish as an animal. In fact, he does not think of all mammals as animals. Whether whales are thought of in this way will no doubt depend on the education of the person concerned. It is certain that *whale* was once believed to be a fish.

The general vocabulary retains the more primitive, the pre-scientific classifications just as the general language retains pre-scientific concepts such as 'The sun rises' (which is what we see with the naked eye). Naturally this creates a certain opacity in utterances and the expression either fails to be self-evident or is misleading. The first sight of a word like *Nilpferd* must suggest a horse rather than a *hippopotamus*. The English word is opaque in the sense that it is constructed from elements of a classical language that is not understood by ordinary people (Greek *hippos* = horse + *potamos* = river). The German hippopotamus is a 'Nile horse' (*Nilpferd*) or a 'river horse' (*Flusspferd*), just as Greek signifies a river horse. This is bad biology but it is deeply rooted in the vocabulary of the given languages. Matters are even worse with a word like *seekoei* (sea-cow) in Afrikaans, which makes for confusion with the Caribbean *manatee*, known in English as *sea-cow*. A hippopotamus is more like a cow in its habits than a horse; it does not gallop but it does browse. The Romance languages tend to go the same way as English in their naming habits, drawing on the classical languages, so that we find F *hippopotame*, I *ippopotamo* and S *hipopotamo*. *Whale* is something of an inconsistency in

English. To be consistent with the Germanic languages, it should be 'whale fish', or should follow the Romance pattern with a Greek-derived name like *baleine* (F). However the Germanic forms are also opaque in many instances; for example, *walrus* = *Walross* ('whale horse'). In the German general vocabulary the *Nilpferd* is a member of the class of *Pferde* (horses), together with the knight in chess (*Pferd*), but not in English.

Technical vocabularies are tidier on the whole than everyday ones. Yet they too vary from the highly systematized and logical to those that are not greatly different from those in general use. The usual procedure is for a generic term to have qualifying elements added to it to express specialized meanings. The qualifier is more often than not an adjective, but it may be a noun or other class of word.

Druck is a general German term for pressure, and also occurs in the everyday language. In the field of mechanical engineering, when the pressure is only in, say, the axially operating direction, it is called *thrust* in English; it has a specialized technical meaning here, distinct from the everyday meaning yet derived from it. The generic term is qualified in the expression *Axialdruck* (adjective + noun). This is a subordinate member of the *Druck* group, in English the *pressure* group. But in English the corresponding term is *thrust*. The German term has the virtue of transparency and logical construction.

But we are not always so fortunate. Take, for instance, *Stoff*, for which the English counterpart is *material* or *substance*. Generality here implies a certain vagueness, just as particularity means precision. A word that includes tends to have more shadowy outlines than the words that it includes. Let us limit the *Stoff* series to a fairly simple sub-group, fuels, for which German has two more specialized terms: *Kraftstoff* (literally 'energy stuff') and *Brennstoff* (literally 'burning stuff'). *Kraftstoff* is a more abstract term than *Brennstoff*, since the concept of 'energy' is a more abstract one than 'burning substance'. On the face of it, *Kraftstoff* should be the including term and *Brennstoff* the more specialized one. In fact it is the other way round. *Kraftstoff* is a type of *Brennstoff*. The English rendering of the first is *motor fuel* and of the second *fuel*. The relationship between generic word and specialized word is much clearer in English than in German on this occasion, and it offers a warning – a warning that can never be repeated often enough in translation – not to take things for granted, and not to assume that things are always what they seem.

It would be a great deal easier if German forms were reproducible by the same structures in English, but the generally logical and systematic German expressions (WW, Ww etc.) do not translate logically into English. Besides *Druck* we have examples like *Mannschaft* from *mann* + –schaft (indicating collectivity) = *team*; *Lastwagen* (load + vehicle) = *truck* (compare *Eisenbahnwagen* = railway carriage, *Krankenwagen* [sick persons = vehicle] = *ambulance*). The German grouping tends to form neat morphological series, whereas the English terms are based on a less transparent system. However the common languages on which the specialized vocabularies are founded all group the lexical items in a more or less arbitrary manner. This is the penalty for the fact that language is not a manufactured object but a system that has developed historically.

The principle of exclusion can be applied to the structuring of vocabulary in the same way as the principle of inclusion. The oppositions already discussed are examples of this. As we have seen, oppositions can be morphological (addition of prefixes like *un-* or suffixes like *-los*, *-less* to the stem) or non-morphological (*good/bad*). The oppositions based on gender are morphological; those based on sex need not be, e.g. *Mann/Frau* (an opposition of both gender and sex); *coq/poule*; *horse/mare*.

Oppositions may be absolute, with one word automatically excluding the other. *Good* is an absolute opposition to *bad*. *Black* is in absolute opposition to *white*. One of the definitions of *black* in the *COD* is 'not white', though this is not, of course, a scientific definition.

Oppositions may also be relative, as with *bigger* and *smaller*, *mountain* and *hill*. Much of the difficulty in translating arises from relative oppositions that are subjective. In most instances *grösser* and *kleiner* will translate into bigger and smaller, but is *Berg* mountain or a *hill*? No doubt *über Berg und Tal* should be rendered as *over hill and dale*, since if *Tal* is in opposition to *Berg* so *dale* is in opposition to *hill* (not *mountain*).

Morphological oppositions again have the limitation of having such a wide range that if the opposition to B is *not-B* (*un-B*), *not-B* may include many items. *Gedankenvoll/gedankenlos*; *thoughtful/ thoughtless* appear to offer an ideal symmetry, but in practice *thoughtful* can be replaced by *considerate* and *thoughtless* by *inconsiderate, unfeeling, insensitive, obtuse* and so on. The adjective *echt* can be translated as *genuine* or *true*, and *unecht* as *untrue*, but there is no word *ungenuine*, and *false* might be the best alternative.

A further principle to be associated with these oppositions is one that might be called reversal, since we have here a pair in which the action or quality expressed in one item is reversed by the other. Here again the reversal may be performed by morphological means, though this is not inevitable.

D	*zulassen*	*zu + lass*	direction towards + allow
E	*admit*	*ad + mit*	direction towards + send
D	*ausstossen*	*aus + stoss*	direction away + push
E	*eject*	*ex + ject*	direction away + throw

It does not matter whether *admit* originally entered English as one form (*admittere*) from Latin or whether it is a combination of Latin elements already incorporated in the language (*ad* and *mittere*). The point is that there is a prefix *ad-* in English that has the sense of direction towards. The same applies to *eject*. The two sets of expressions correspond in regard to direction (towards/away), but *lass* and *mit* are not perfectly equivalent, since the latter is more specific. Both *stoss* and *ject* suggest forcible action.

Non-morphological reversal is expressed in the pairs:

kommen/gehen	*come/go*
nehmen/geben	*take/give*
kaufen/verkaufen	*buy/sell*
borgen/leihen	*borrow/lend*

(In the German pair *kaufen/verkaufen*, the reversal is of course morphological, but not in the corresponding English pair.)

In this list the first item in the pair expresses direction towards the speaker, the second direction away. A different type of reversal is displayed in the following set:

aufbauend	*auf + bau*	direction up + build
zerstörend	*zer + stör*	into pieces + disturb
constructive	*con + struct*	with + build
destructive	*de + struct*	remove + build

A non-morphological contrast is provided by *sitzen/stehen*: *sit/stand*. The concept is of something static, but 'standing' undoubtedly reverses 'sitting'.

Pairs like *gesund/krank*: *well/ill* present a difficulty in that they may suggest alternative parallels (*gesund/krank*: *healthy/invalid*), depending on whether a present state or a general state is intended. The answer to the question: 'How is he?' is 'He is well', not 'He is healthy'. (He is well at the present time; he is healthy all the time.)

Another opposition occurs when two things are contrasted that are not direct opposites but stand in opposition to one another. True opposition is expressed in *reich/arm*: *rich/poor* (presence of means/absence of means). There is opposition of a different kind in *reich/mittellos*: *rich/destitute* (presence of means/absence of means). *Mittellos*, *destitute* suggest an extreme of poverty beyond that indicated by *arm*, *poor*. In the first series (*rich/poor*), more general terms are used, which in turn are members of a more general class. The *rich* is the class of the *rich* (animate), and the *poor* likewise is *poor* (animate), applicable to human beings. *Destitute* is a member of this class, together with words like *needy*, *indigent* and so on.

Words can also be classified in terms of concepts like animate/ inanimate, static/dynamic (stand, run), natural/artificial, and a number of other categories.

In addition to these contrasting types, systems can be established in regard to degrees of difference and similarity. Something of this nature is contained in *bigger/smaller*, though what is intended here is scale. Scales may be logically constructed and may have either a quantitative basis or a subjective basis that is systematic and controlled, as with organoleptic tests in food technology and various psychological tests. In the ordinary language grading is applied to many objects of everyday experience.

A good example here is the distinction between *mountain* and *hill* or between *forest* and *wood*. *Mountain*, according to the *COD*, is 'Natural elevation of earth's surface, large or high hill', and *hill* is: 'Natural elevation of earth's surface, small mountain'. A similar distinction is made between F *colline* and *montagne* and D *Berg* and *Hügel*. The one is defined in relation to the other, and for the layman it would be difficult to say exactly when a hill becomes a mountain or vice versa. Northern India's hill may be Australia's mountain. In German *Forst* is a cultivated forest (*Wirtschaftswald*), as against *Wald*, but in English a *wood* is considered to be smaller than a *forest*. Similarly, it is not easy to distinguish between a town and a city in terms of size. To most people a city will be a large town and a town a small city, but ideas on this will vary according to circumstances. A German *Stadt* is either a town or a city, but if the speaker wishes to emphasize that it is a city in the general English sense of a large town, he will have to use *Grossstadt*.

Series of words shading off into one another in this way are used for many objects in the environment, but in each language the grading is distinctive and peculiar to itself. This is exemplified by some of the expressions used for running water:

D

Strom	large or broad river, stream; current (also of electricity)
Fluss	river, stream; flow, flux, melting, fusion; molten glass or metals
Bach	brook, stream, rivulet, rill

E

river	copious stream of water flowing in channel to sea or lake or marsh or another river (*COD*)
stream	body of water running in bed, river or brook (*COD*)
current	running stream; water, air, etc, moving in given direction (*COD*) (also used in electricity)

F

fleuve	important watercourse emptying into the sea
rivière	watercourse that joins a more important one
ruisseau	small watercourse
courant	stream, current (including electric current)

It is clear that the senses crisscross and overlap in a disorderly manner. German *Strom* is a somewhat archaic expression that has acquired a modern usage in electricity, corresponding to E *current* and F *courant*. German *Fluss* has a range of meanings extending beyond *river* or *stream*. The French series is precise and systematic.

Some systems of relationships may be peculiar to one language system. A striking example is kinship. It is well known that kinship systems do not coincide perfectly, even between closely related languages. For example there is close contact between Afrikaans-speaking and English-speaking groups in South Africa; they share many values and interact daily, yet their kinship terms deviate; (and the deviation from those of the more distant Bantu languages is even greater).

D	E	Af	Sv	
Grossvater	*grandfather*	*grootvader/ oupa*	*farfar/morfar*	(father's father/mother's father)
Grossmutter	*grandmother*	*grootmoeder/ ouma*	*farmor/ mormor*	(father's mother/mother's mother)
Vater	*father*	*vader/pa*	*far*	
Mutter	*mother*	*moeder/ma*	*mor*	
Sohn	*son*	*seun*	*son*	
Tochter	*daughter*	*dogter*	*dotter*	
Onkel	*uncle*	*oom*	*farbror/ morbror*	
Tante	*aunt*	*tante*	*faster/moster*	(father's brother/mother's brother)
Schwager	*brother-in-law*	*swaer*	*svåger*	
Schwägerin	*sister-in-law*	*skoonsuster*	*svägerska*	
Bruder	*brother*	*broer*	*bror*	
Schwester	*sister*	*suster*	*syster*	
Geschwister	*brothers and sisters*			
Gebrüder	*brothers*	*gebroeder*		
Neffe	*nephew*	*neef/broers-kind/suster-kind*	*brorson/ systerson*	(brother's son/sister's son)
Nichte	*niece*	*niggie/broers-kind/suster-kind*	*brorsdotter/ systersdotter*	

German, English and Afrikaans tend to classify relationships in terms of the succeeding generations, whereas Swedish also has a (vertical) classification on the father's and mother's side. This extends to grandchildren, who are *barnbarn*, but also *sonson*, *dotterson*, *sondotter*, *dotterdotter* and so on.

Time systems also vary between languages. The day is divided up differently in German, French, English and Swedish. Swedish *dygn* lasts twenty-four hours (day and night) German *Morgen* and *Vormittag* correspond to French *matin* and English *morning*. *Mittag* is equivalent to *midi* and *noon*, but also to *après-midi* and *afternoon*. *Nachmittag* corresponds partly to *afternoon* and *après-midi*; partly to *evening* and *soir*. *Abend* can also be treated as *evening*. In Swedish you can say *God Morgon* up to 10 am, thereafter *God Dag*.[4]

Temperature differentiations form another area of relationships. Temperature can be measured quantitatively by instruments and the exact degree can be expressed according to a scale. The temperature words used in the general language, however, are extremely subjective and the same expression, *heiss* (*hot*) or *kalt* (*cold*), can vary according to the situation in which it is used: climate, liquids, food and so on. Midway between *heiss* and *kalt* comes *warm*. But the position of *warm* in the German system of temperature sensations does not correspond to that of *warm* in the English. The German expression can designate higher temperatures than the English one. *Eine warme Suppe* is soup at the right temperature, but *warm soup* is not hot enough, suggesting tepidity. *Warmwasserversorgung* is *hot water supply*, not a supply of warm water. Similarly *heiss* tends to range towards a higher temperature than *hot*. *Torrid zone* is the English equivalent of *heisse Zone*. The range of *kalt* does not fully correspond to that of *cold* either. Again, if we take a geographical example, the German extends to a more extreme point than the English, with *kalte Zone* for *frigid zone* and *mich friert* (I freeze) for *I am cold*.

Certain restrictions in the use of these adjectives should also be borne in mind. *Kühl* (*cool*), for instance, is used only to express climatic conditions and liquids (especially beverages). Similarly *siedend-heiss* is used only with liquids and thus corresponds to *boiling hot*, but it overlaps with *kochend heiss*, which can be used of both liquids and foods.

The range of colour terms in many languages has been studied, and eleven basic categories have been found from which eleven or fewer basic terms are always drawn: white, black, red, green, yellow, blue, brown, purple, pink, orange and grey. The following points appear to have been established:

1 all languages contain terms for white and black
2 if a language contains three terms, it will also have a term for red
3 if it contains four terms, it will include either green or yellow
4 if it has five terms it will have green and yellow
5 if it has six terms, it will have blue
6 if it has seven terms, it will have brown
7 if it has eight or more terms it will include purple, pink, orange and grey, or a combination of these.[5]

But of 2048 possible combinations of basic terms, only 22 occur in practice. This study is extremely suggestive and points a way to similar studies that may be of value in investigating lexical structure.

From this list, however, it is evident that all these universal categories do not exist in every language at the same time, since individual languages vary in the number of categories they contain. In addition, the range of a colour term in a language is not necessarily the same as that of a similar term in another language. There is a difference between the colour system used in optics and that used in chemistry, in sales language (fashion words) or in the everyday language. There will be greater correspondence between scientific usage in different languages than in the everyday language, or the language of, say, cosmetics. In German, for instance, words like *orange* and *violett* are of fairly recent origin, and the range now occupied by *orange* overlaps with that of *rot* (*red*). Orange used to be designated by *rotgelb* (reddish yellow) and *violett* by *blaurot* (bluish red). In scientific uses the terms have more precise applications and should not give rise to difficulty, e.g. *Ultraviolett, ultraviolet*: *Ultrarot, infra-red*.

MULTIPLE MEANING

Multiple meaning (polysemy) assumes a number of forms. Words may have the same pronunciation but different written forms, as in *right, write, rite*. Translation, being involved with the written language, is not concerned with this so much as with the type of homonymy (same name, different meaning) in which the visual form has more than one meaning (homography). The German gender sytem makes it possible to distinguish between different homographs such as *die See* (*the sea*), *der See* (*the lake*); *das Tor* (*the gate*) and *der Tor* (*the fool*). But this does not apply everywhere. For instance, *der Laden* may be either *the window-shutter* or *the shop*. Even in technical fields, homography occurs. It is particularly evident across subject fields. An expression like *plasma* is used in biology, physics and mineralogy, and in each of these it has a distinctive meaning. *Axes* in English may be the equivalent of *Achsen* (pl. of *axis*) or *Axte* (pl. of *axe*); *concrete* that of *konkret* (as opposed to *abstract*) or *Beton* (construction material).

Polysemy also results from expressions that are common in the general language as well as in the special languages. This large group of words includes *Arbeit, work* (physics, engineering) and *Kraft, force* (physics), which in everyday language are extremely vague and have a wide range of meaning. The difficulty can sometimes be resolved by the context, though not always. The TL equivalent may be acceptable at a pinch without being as precise as it should be. This is a barrier to communication. An example is the word *Geschwindigkeit*, which sometimes means *speed*, sometimes *rate* and sometimes *velocity*. The use of any of these may be intelligible but it may result in irritation on the part of the reader and loss of time.

Polysemy is also found across languages when, as with *Geschwindigkeit*, the expression has several alternative renderings in the TL. This does not present too much of a problem, since the alternatives, as discussed above, are related to each other in meaning. But let us take F *champignon*, which can mean E *mushroom*, *toadstool* or *fungus*. Although the English words are related, it is possible to imagine a situation where a lack of precision in translation might have serious consequences.

In addition to multiple meaning of words that have the same written form, words that are nearly the same should also be mentioned. Examples include *der Muff* (muff) and *die Müffe* (coupling); *die Akte* (documents, proceedings of a learned society) and *der Akt* (action, act in a play); *die Etikette* (etiquette) and *das Etiket* (label).

Important here are the one-to-two correspondences between the SL and the TL. The SL word has multiple meaning as far as the TL is concerned. A case in point is *Falte*, which in German can be used of either a piece of cloth or of skin, whereas in English it must be *fold* for cloth, *wrinkle* for skin. Similarly *Ende* in German refers to sharp or blunt, hard or soft objects, and there are three possible equivalents in English: *end* (termination) in the most general sense; *tip* ($\pm$ blunt, $\pm$ soft, but not hard or sharp); point (sharp $\pm$ hard, not blunt).[6]

Synonymy can also present problems in technical literature. This is especially true of medicine. *Terapéutica tópica* is the Spanish for either *topical therapy* or *local therapy* in English. *Rubeola* is in many languages *German measles*, but it can designate either *German measles* or *measles* in English. It is estimated that between a quarter and a half of medical and paramedical terminology consists of obsolete, rare, officially rejected or otherwise undesirable synonyms. More than thirty terms have been used in English, and as many – though not necessarily similar ones – in German, to describe 'myelofibrosis'.

The expressions *Beständigkeit*, *Festigkeit*, *Resistenz* and *Widerstand* are all translatable as *resistance*, while in electrical terminology

Widerstand is *resistor*. Then in chemistry and pharmaceutics, we have the example of Na_2SO_4 (sodium sulphate, commonly known as Glauber salts). In chemistry the German expression is *Natrium-sulfat*; in pharmacy *Schwefelsaures Natron* or *Natrium sulphuricum*; in popular speech, *Glaubersalz*.

Another example of a term that can be troublesome is *Röntgeno-metrische Bestimmung* and *Röntgenometrische Auswertung*, both of which mean *X-ray analysis*. If the translator assumes, as is only natural, that they must each have a specific and distinctive meaning, a great deal of his time can be wasted.

RANGE OF COLLOCATION

Certain combinations of words are not fixed by means of grammatical rules but as a result of their meaning and long association. Some of these words can be said to have a tendency to combine with certain other words; in the case of others we are restricted in the choice of words with which they may be associated. These tendencies or restrictions may apply in so many different circumstances that they can be referred to as general rules, or they may apply in specific circumstances only. The word *ream*, for instance, can be used as a noun only when qualified by *printer's* or in the expression *a ream of paper*.

The general rules are sometimes incorporated in grammar books, including the restrictions on animate and inanimate verbs as far as their subjects and complements are concerned. The verb *see* must have an animate subject, except when it is used in a derived sense, as in a metaphor. Other restrictions classify the word as static or dynamic: *stehen* (stand, stay) and *gehen* (go). These are inherent restrictions; they are as it were built into the word as it is used in the particular language.

Other combinations may be merely the result of habit. We say *lightning strikes*, rather than *lightning smashes* or *lightning cuts* or *lightning hits*. The sentence 'Lightning hits the roof' is neither unintelligible nor ungrammatical; it is indeed acceptable, but it is not a standard expression. Such combinations are stylized, fixed, stereo-typed.

Preferences of this kind naturally vary from language to language. On the whole they vary far more than grammatical restrictions, and restrictions governed by meaning. There is nothing in the meaning of *lightning* that makes it obligatory to use *strike* with it instead of, say, *thrust*. The only requirement is that *lightning* must have a dynamic verb. It cannot have an animate verb, since it does not see, talk or hear, but it can strike, hit, flash or leap from cloud to cloud. German *der Blitz* has many of the vivacious attributes of English lightning. *Es blitzt* means not only *it lightens* but also *it flashes, it sparkles*.

Restrictions based on a conceptual affinity are of a different order. An action like *beissen* (bite) presupposes *Zähnen* (teeth). Teeth alone can bite, unless the meaning is extended by metaphor. The original concept may remain with some of the derived expressions, in which case a good correspondence between languages may be expected. The counterpart of *beissender Scherz* is *biting wit* (wit with teeth in it) and though *ins Gras beissen* (**bite into the grass*) is not quite *bite the dust*, the relationship is still visible. But *es beisst mich in die Augen* (**it bites me in the eyes*) is unknown in English, which has the less violent equivalent *it makes my eyes smart*. Similarly, *der Spaten* (*spade*) is conceptually associated with *graben* (*dig*), and *graben* tends to have as complement *Loch* (*hole*). Spades are thus restricted in their combination with verbs in German; although a spade can be used for striking and many other things, its most usual function is for digging, and digging normally entails making a hole in the ground.

The two extremes are a virtually unrestricted capacity to combine, and restriction to only one or two words (*ream*). An example of free combination is the adjective *wet*, since we can talk of *wet sand* as against *wet floors*, *wet clothes*. Similarly *dry sand*, *dry floors*, *dry clothes* are all used. But the contrast to a metaphorical *wet blanket* is not a **dry blanket*, and the contrast to *dry humour* is not **moist humour* or **humid humour*. Similarly, *dry humour* is not equivalent to **arid humour* and *wet blanket* is not identical in meaning with **moist blanket*.

Green board offers an example of a loose combination. Other colours can easily be substituted for green and other objects for board. The meaning of the phrase is very largely the sum of its parts. But we cannot substitute *iron* in *lead pencil* without producing a totally different idea. An *iron pencil* would be a pencil made of iron. A *lead pencil* is not a pencil made of lead but a pencil that contains *a lead*, which is not the metal lead but, usually, graphite. As long as the word *lead* is juxtaposed with *pencil*, it conveys this meaning; the moment it is attached to another word, it becomes simply the metal, as in *lead box*. The same applies of course to pencil. *Lead pencil* is a single lexical unit, the result of long association between the two words. In German they are compounded into one word, *Bleistift*. But in other contexts *Blei* means the metal lead or, at any rate, not graphite or pencil.

A large number of fixed combinations recur frequently in technical language. Some of these are modelled on officialese, which is very fond of stereotyped phraseology.

The expressions *Erfolg haben* (*have success*) and *erfolgreich sein* (*be successful*) are best translated by *succeed*. Similarly *Anordnungen treffen* (*make arrangements*) means *arrange*, *Bericht erstatten* (*make a report*) means *report*. But the more literal translations in parentheses are just as common in English. Another common grouping in technical phraseology is *Arbeit . . . durchführen* (*carry through work*), or *carry out work*. The phrase *der Stand*, like *the state of*, is likely to elicit *der Stand der Technik* = *the state of the art*. Other well-worn associations are *von besonderer Bedeutung* = *of special significance* and *the order of magnitude* (*die Grössenordnung*). The last example illustrates how a collocation in one language can be a compound in the other, confirming the closeness of the bond. But pairs with a magnetic attraction for one another in one language may be incompatible in the other. Thus there is a temptation to render *erheblich gross* as **considerably great*, on the pattern of *erheblich grösser*, which has a mirror image in English *considerably greater*. But considerably cannot collocate with *great* in English, so the translation must be *quite large*, or *very big*, or something of that kind. Similarly it would be incorrect to translate:

1a *Die Messungen dieser Grössen bereiten oft erhebliche Schwierigkeiten*

as:

1b **Measurements of these factors are often considerably difficult*

Here again the correct expression is *very difficult*, or *extremely difficult*, though 'often causes considerable difficulties' would be correct. Alternatives would be: *significantly difficult* or *tremendously difficult*.

Other adjectives tend to be restricted to animates or to inanimates. The adjective *dick* can apply to anything, but *thick* is normally used for inanimate objects, while *fat* and *plump* apply to living creatures and *buxom* is used only for women. *Gross* has general application but *hoch* and its counterpart *high* tend to be restricted to inanimates. *Tall* is mostly used with human beings but can also be applied to certain inanimates such as trees or furniture. *Alt* and *old* are unrestricted, but *bejahrt* can be used only for human beings, while *aged* is restricted to people and certain animals (such as horses). *Weit* is restricted to hollow spaces like tubes and surfaces, whereas *wide* can be applied to solid objects as well. A complicating factor is that the range of application of an adjective may be extended by its use in a metaphor.

Situation

𝕊𝕊𝕊𝕊𝕊𝕊

THE communicative situation embraces all the circumstances in which a message occurs. It embraces the grammatical features of the languages concerned, the properties of their vocabularies, the subject field of the text, the author of the text and his intention, the reader and his interpretation, the time, the place and various other matters.[1] The simplest form of the communicative situation is: 'Who says what to whom?' There is a message source (author) (W), a message (M), a channel (language used) and the receiver of the message (R). In addition we have a translator (T) and a second channel. The original channel is the sL and the second the TL. Other influences operate in the chain besides those mentioned, and in view of these many links, perfect communication is an ideal rather than a practical possibility. What is achieved is an approximation, which is more or less close to what the author (W) intended. What is put in at W is not equal to what comes out at R, but it may be near enough to meet the requirements of the situation. It follows from this that perfect translation, too, is an impossibility, but that we can reasonably hope to achieve a translation that will satisfy practical needs. These are the assumptions underlying all technical translating. The significance of this way of expressing the situation is that it obviates a discussion on whether translation is possible at all, and instead places the emphasis on attempting to achieve the highest possible degree of correspondence.

We can concentrate on any of the links in the chain of communication: author, reader, message. The one link to which least attention is paid is – paradoxically enough – the translator, because the ideal technical translator should be as unobtrusive as possible. A successful translation is one that appears to have only three stages: author, message and reader. It should not read like a translation at all. Some remarks will be made later about the personality of the translator, but we can forget about him for the moment.

In technical translation the emphasis is on the effect produced on the reader. The technical writer's manipulation of the language is always exercised in terms of this effect. He is not seeking self-expression in the way that a poet does. His self-expression has already been achieved in the work he is describing. The written document is not an end in itself, as a poem or novel is, but merely a means. The author's intention is to communicate certain facts or theories to his readers. There is, however, a type of technical text in which different motives are at play. Patent literature, for example, requires the author to communicate as much information as is required by the patent legislation, but no more, and to conceal as much information as possible. This gives rise to a special style, and this style is additionally affected by the need to comply with certain statutory and legal requirements. In sales literature, on the other hand, the intention is to persuade the reader. This may be achieved by means of facts and theories, or by emotional appeals, or by a mixture of the two. This process again results in a special style. Even where the author intends to communicate fact or theory alone, he may need to vary his style according to the type of reader (a research worker or engineer applying knowledge; a reader active in pure science or in technology; and so on). Within the field of technical prose there are a number of sub-species, each with its own specific customs and conventions concerning what should be said and the way to say it. Peculiarities of vocabulary also exist, plus the occasional grammatical peculiarity. At the same time we must not forget that any style is part of the language as a whole, that it shares its general nature, that its vocabulary and any grammatical peculiarities it may possess are automatically part of the language.

It is indispensable in technical communication for both author and reader to know the subject field. Each language variety and each subject field creates its own expectancies. In the case of the word *Dämpfschnecke*, the reader will be guided by his knowledge that it occurs in a particular variety of technical prose. Thus a reader who thinks he is going to be told something about biology will try to interpret the term in a different way from an engineer. He will expect it to refer to some kind of snail, whereas a reader expecting a statement on engineering will interpret it as a worm.

The terminological series of compounds in which *Schnecke* is the main element would tend to have an equivalent in an English series using *worm*, as in *worm gear*. Two points should be noted here: the subject field determines the selection of one meaning (ideally speaking) from a number of others; secondly, the fact that the object denoted is called a *Schnecke* (snail) in German and a *worm* in English. This relates to a phenomenon that has been mentioned earlier: the aquatic African pachyderm known in English as a hippopotamus is a *Nilpferd* (Nile horse) in German, a *flodhäst* (river horse) in Swedish and a *seekoei* (sea cow) in Afrikaans.

Every language system harbours items of this kind, words that denote an object in the external world in accordance with certain beliefs now obsolete. If the native speaker is aware of this, he may consider it a curiosity. He knows that a hippopotamus is not a kind of horse and calling it one will not make him believe it is. From this point of view, the vision contained in the word has become blurred. The word is a mere label. This is true of many words in the language, but for many words it does not apply. The native speaker is not conscious of any incongruity between the designation and what is denoted. In that event, two things are possible: it may have little influence on his thinking, or it may have a significant influence. We talk of the sun rising, yet every literate person today knows enough astronomy to know that the earth moves in an orbit round the sun. This seems contradictory, but in fact our naked eye tells us that the sun rises. It is perfectly possible to think of the sun rising each day and at the same time to think of the earth in orbit round the sun, for these two ideas occur within different frameworks of knowledge. We live constantly with such separate frameworks. If speaking of 'the sun rising' were to influence an astronomer into literally believing that it does, that would be another matter. But there is no evidence to indicate this. In some spheres language does undoubtedly tend to predispose the speaker to think along certain lines: where words have an emotional loading, or play a role in propaganda, for instance, there is no doubt that built-in associations can be played on.

We could also cite numerous instances where no ideological or propagandistic application comes into play but the expression will still have an influence. Thus an untutored language user coming across the phrase for the first time might be forgiven for assuming that a Tasmanian wolf is just that – a wolf whose habitat is Tasmania. To the biologist, it is not a wolf at all. To the inhabitants of Tasmania, their marsupial wolf is what the true mammalian wolf is to the inhabitants of Europe. There are no wolves in South Africa either, and yet the Afrikaans language has words like *Strandwolf* (*beach wolf*) or *brown hyena*. South Africa has no tiger, and yet a *tier* can be either *tiger* or *leopard*, an animal that is found there. In the same way South American Spanish uses *tigre* for *jaguar*, as there are no tigers there either. The denotative range of *tiger* is wider in these two instances than it is in English – it refers to two animals instead of one.

Other examples illustrate an extension of meaning whereby existing words are used to describe new phenomena. These phenomena resemble others designated in the language, so the name is transferred. The old ways of classifying nature persist in the new country. We thus have both a continuity with the past and at the same time a branching off.

The question of continuity is extremely important. If we associate with it the presence of common values, common habits, traditions and so on we have a phenomenon that has been called 'cultural overlap'. And, as Lyons says, 'It is precisely the fact of considerable cultural overlap, deriving largely from the continuity of the Greco-Roman and Christian cultural norms and institutions, which makes translation between the different Western European languages relatively easy and fosters the view that the distinctions made in these languages are absolute distinctions given "in nature".'[2]

The 'river horse' is an example of this principle. The same metaphor runs through a number of languages, bearing evidence of cultural contact between a number of languages (German, Swedish, Dutch, among others) in this translated form. The *hippopotamus* loan is similarly found in another range of languages (French, Italian, Spanish, English). All of them have derived it via Latin from Greek.

Schnecke follows the same pattern. In the languages of Western Europe we can demarcate families of images for this notion. We could start with Archimedes' screw, where the similarity to the pattern of the mollusc shell gave rise to the use of the expression *helix* (snail, spiral shell). The 'snail' (spiral shell) notion is found in

German *Schnecke* and has influenced languages like Swedish (*snäck/a*) and other Scandinavian languages, the Slavonic languages, Italian *coclea*, Hungarian *csiga* and so on. As against this, we have a 'worm' family of languages, which includes English, Dutch (*worm*), Afrikaans (*wurm*) and Japanese *womu*. We also have a 'third world' of 'neutral' or 'geometrical' cast where the model is not biological but abstract, e.g. French *vis* (screw), Rumanian, Spanish (*rosca*).

Alternative forms are also available: English *worm* is replaceable by the more abstract *screw* or *spiral*; and alongside French *vis* is *hèlice* (from Greek), though the average French speaker will think of it as an abstract term, with no connection with *escargot* or *limace*.

If we know that the 'snail' image derives from the ancient Greek 'water snail', the original Archimedes' screw, this explains the origin of the 'snail' cultural sphere, though not fully its spread and influence. The translator is always concerned with current usage. The native speaker uses the term in its current meaning, not in its original one. A French speaker, as already observed, scarcely thinks of *hélice* as having anything to do with a snail. But where the concrete word 'snail' (as in German), 'worm' (as in English) is used, the image exercises a subtle influence. Much technical terminology and nomenclature is coined from the classical languages, but not all. It is derived to a large extent – perhaps to a greater extent than is generally realized – from the raw material available in the ordinary colloquial language, as expressions such as *soup* and *breeder reactor* testify.

If 'snail' originally entered technical language as the result of a physical similarity between a mechanical device used for lifting water and the shell of a mollusc, once the term existed in the vocabulary with this meaning, the extension of that meaning was inevitable.

For the translator, the effect of this process is twofold. It facilitates correspondence on the lexical level between languages in the 'snail' world. The terminological series is not consistent, however, and the *Schnecke* compounds do not all represent the same selection of features from the snail shell. A shell can suggest not only a spiral shape but also housing (as compared to the lack of housing in a slug). Inconsistency is also found in the *worm* world. This leads to unpredictable divergences:

D	E
Schnecke (*Transportelement*) (*transportation element)	*worm*
Schnecke, 2-teilige Paddel (*2-part paddle)	*crescent blade worm*
Schnecke, 4-teilige Paddel (*4-part paddle)	*paddle-blade conveyor*
Schnecke (Maschinenelement) (*mechanical element)	*worm*
Schneckennetzapparat (*snail moistening apparatus)	*damping worm*
Schneckenrad (*snail gear)	*worm wheel, worm gear*
Schneckentrieur (*snail trieur)	*spiral trieur*
Schneckentrog (*snail trough)	*trough casing*

Expressions of this type should be referred to the object denoted in the external world. In principle, they can be identified in this way. It is harder to deal with more abstract expressions.

Gewerbe, for instance, can be interpreted as trade, business, calling, profession, industry, vocation, occupation. An expression such as *die Müllerei* would be designated in a monolingual dictionary as *das Mühlengewerbe* and in a bilingual one as 'the miller's trade' or 'the miller's trade or craft'. In German milling is regarded as a *Gewerbe* and one of the designations of *Gewerbe* is *trade*. This probably corresponds to the general view. A German speaker will think of 'milling' as a *Gewerbe*, but not the plastic and chemical industries, since *Gewerbe* has the implication of *Handwerk (craft, trade)*, i.e. an older stage of technology, as against *Industrie*, a modern stage. It is associated with archaic as against up-to-date methods. Milling is referred to as a trade in English only when the speaker deliberately chooses to emphasize its antiquity. The most common term is *milling* or *the milling industry*. In a contemporary framework *die Müllerei* undoubtedly denotes a modern, up-to-date enterprise. Yet it is often translated as *the miller's trade* or *the milling trade*. This is translating in terms of the dictionary and designative value rather than taking into account the denotative range of the term, i.e. what it can refer to in the external world. By translating only on the lexical level we submit to the obsolescence and inadequacy of dictionaries, particularly bilingual ones. The translator is trapped by words and has ignored the external situation.

But even in terms of words, thanks to the differences in the vocabulary structures of German and English, *Gewerbe* has a wider range than *industry* and a different range from *trade*. It overlaps to a certain extent with both words, but not entirely. This means that *Gewerbe* can in some circumstances be translated as either *trade* or *industry*. In the field of food technology, for instance, *das Gastgewerbe* (the hospitality trade) translates in a phrase like *Detail – als auch im Engrosverkauf (einschliesslich des Gastgewerbe)* as *the catering industry: retail and wholesale trade (including the catering industry)*.

To German speakers milling is still run on craft lines, whereas in English *milling* tends to collocate with *industry*. This is no doubt because industrialization took place earlier in Britain than in Germany and the craft tradition was largely obliterated, whereas in Germany the craft tradition is still a vital force.

This is an important point to remember when you are translating from German into English, for any attempt to produce an exact translation of *Gewerbe* will show that it is a precipitation of the medieval guild tradition in the language.

It is clearly necessary to understand the concept underlying, say, *die Müllerei, Gewerbe* and the *-schnecke* series before we can arrive at the appropriate English equivalent. The dictionary will not give it to us and indeed may mislead us badly, since the difficulties in translating such expressions are due to cultural divergence. The *Schnecke-worm* contrast is one expression of differences over a wide area of experience, and can be summed up as differences in cultural range.

For our present purposes culture can be classified as follows:

ecology	physical environment; climate; topography; flora; fauna
social relations	institutions; educational system; social stratification; politics; historical experience; economics
technology	general level of material culture; state of science and technical knowledge; industry, agriculture
beliefs and values	political and religious beliefs and traditions; values

language and art type of language (flexional/non-flexional; grammatical categories, syntax, vocabulary); dialects and regional variations; aesthetic background

Underlying all this is the universality of our biological and physiological make-up. All human beings share certain needs and drives – eating, sleeping, breathing and so on. In addition we have certain psychological attributes in common. These universals help to make translation possible, whereas cultural peculiarities make it a difficult task, though the sharing within Western Europe of certain common traditions outweighs this to a certain extent.

For the technical translator it may not be important that the Eskimos have a profusion of lexical items to describe different types and states of snow between which a West European language would not discriminate. Yet ecological conditions do affect a German's attitude to forests and mountains, not only in regard to emotional associations, but also in determining his means of distinguishing between, say, a mountain and a hill.

Differences in institutional life have an effect on the translatability of expressions such as *Gewerbe*, *Anstalt*, *Hochschule*, academic titles and so on. Beliefs and values exercise a very profound and very subtle influence on perception and linguistic expression, among other things on idiom. Even biological and physiological universals are culturally tinged. Everybody eats, and yet dietary habits vary immensely. Foods differ, their manner of preparation differs and habits of consumption vary. Corn may be the staple dietary item in the West; in the East it is rice. In the West knives and forks are used; in the East chopsticks. These are obvious contrasts, but there are also differences between German and English habits.

There are also differences in the names of institutes and committees of various kinds. The DNA (*Deutscher Normenausschuss*) is translated to correspond to its nearest English-language equivalent in function and so is rendered *German Bureau of Standards* or *West German Bureau of Standards*. The acronym DNA may be retained thus: *DNA* (*West German Bureau of Standards*). However, the standards issued by the organization *Deutsche Industrie-Normen* are not rendered as *German Industry Standards*, but with the abbreviation *DIN* 1000, since this designation constitutes a reference.

Thus *Änderung gegenuber DIN* 10 000 should be translated as *modification of DIN* 10 000. The phrase *Deutsche Normen* is rendered as *German Standards*.

In case of doubt, it is always wise to leave the name of the organization in the SL. Where the organization is well known there will generally be a well-established English equivalent. It is also possible to give the original name together with a TL version in parentheses: *Deutsche Gesellschaft für Luft- und Raumfahrt (German Aeronautical and Astronautical Society)*. Where there is an abbreviation, the practice with DNA can be followed: *Der Ausschuss für Lieferbedingungen und Gütesicherung (RAL) beim Deutschen Normenausschuss* = *The Committee for Conditions of Supply and Quality Promotion (RAL) of the German Bureau of Standards (DNA)*.

Similar improvisation and approximation is necessary for a range of organizations and relationships that have no exact English counterpart:

Bundesverband der Deutschen Industrie	*Federal Association of German Industries*
Zentralverband des Deutscher Handwerks und Deutscher Handwerkskammertag	*Central Federation of German Trades*
Hauptgemeinschaft des Deutschen Einzelhandels	*Board of German Retailers*
Deutscher Städtetag	*German Urban Association*
Deutsche Angestellten-Gewerkschaft	*German Trade Union of Salaried Employees*
Arbeitsgemeinschaft der Verbraucherverbände	*Institute of German Consumer Organizations*
Bundesministerium für Wirtschaft	*Federal Department of Economic Affairs*
RKW (Rationalisierungs-Kuratorium der Deutschen Wirtschaft)	*RKW (Rationalization Board of German Commerce and Industry)*

Peculiarities of history and social conditions give rise to ideas that are so closely bound up with a particular culture and language that it is difficult to provide anything more than a vague correspondence in the TL. Such an expression is *die Ernährungsreform*. It presents no grammatical problem: it is a WW type of compound, and the elements can be deciphered (*the nutrition reform). But what does it mean? A concept has been institutionalized in German, but not in English. 'Nutrition reform' is ambiguous: 'reform' does not normally occur in association with 'nutrition' in English, but with organizations and institutions. 'Nutrition' is also liable to be found in a technical vocabulary in English, rather than in everyday contexts. The term therefore conveys information about food, about diet and about nutritional reform, the latter suggesting the realm of public affairs rather than private decision. Politics and diet? In the past dietary movements were considered as cranky in English-speaking countries. The whole idea of a 'movement' concerning food has an alien ring. This sort of thing can be said in English only by means of circumlocution. Who would carry out the reform? A legislature? What would it consist of? The nearest equivalent might be *natural food movement*, which sounds less organized than the German and less pompous, and has less of a taint of faddism. The sense is that of a change to healthier dietary habits, but even expressing it in this way can never give the exact meaning of the German. The very compactness of the term in German is a sign of its naturalness in that language.

This example also raises the problem of different levels of discourse: an elevated style or terminology in one language does not appear so in another. In this sense translation tends to produce a stepping up or stepping down. German, including technical German, is on the whole more formal than English. A technical phrase like, say, *die verwendete Elektrizitätswerkflugasche* again illustrates the same phenomenon – partly because of the German capacity to form elongated compounds. English *the utilized power station fly ash* looks a great deal simpler.

This sort of thing does not happen only between German and English. It also occurs when a Romance language is the SL. The terms used in Italian sound extremely high-flown from the English point of view. Thus *l'ingegneria sanitaria* in a fairly straightforward context must be translated not as *sanitary engineering*, but as *sewage works*. Compare also:

I *tali tipi litologici*	(*such lithological types*)
	these rock types
I *uno strato di terreno agrario*	(*a stratum of agrarian terrain*)
	a stratum of agricultural land

D *die maximalen Werte der*	(*_the maximum values of the_
Spannung und des Stromes	_voltage and of the current_)
	under maximum voltage and
	current

A more literal translation would undoubtedly be a hindrance to easy understanding of the text, though it might not obscure the meaning.

Occasionally the position is reversed and the English offers a higher level of abstraction:

| D *Trockendruckverfahren* | (*_dry printing process_) |
| | *Xerography* |

Differences in economic life also leave their imprint. German distinguishes between *Grossbauer*, a farmer with large land holdings, a wealthy peasant (possibly to borrow from Russian, *kulak* would be nearer the mark); *Bauer*, a peasant, a small farmer; and *Kleinbauer*, a smallholder. But, significantly, it has also borrowed the word *Farmer* from English to designate farmers in America i.e. colonial settlers.

Equally significant is a word like *Landeskunde*, which means both *regional geography*, a concept readily understood in English, and also the idea of scientific knowledge and presentation of one's own country, which has a folklore tinge. *Stammland*, too, is very hard to convey in English. The usual translation is *country of origin*, but *Stamm* has a wide range of associations covering race, clan, tribe, stem, stock, breed, strain and so on.

Terms used in addresses can also present difficulties. What is the difference between a street, a road, a lane, a crescent or a square? However, this does not in practice present a serious problem, since as a rule a street address will be left unchanged, e.g. *Wilhelmstrasse* 111.

National and cultural distinctiveness can also be found in well-known technical terms. It has been observed that English talks in terms of *voltage*, German of *Spannung* (tension). French prefers to talk in terms of current, *intensité*. Names of equipment may vary. The Germans call X-rays, *Röntgenstrahlen*, and *Röntgenbild* = X-ray *photograph* = F *radiographie*. Many terms deriving from the names of discoverers are international: *henry*, *hertz*, *watt*. These no longer strike us as names but as units of measurement. However, the national designations do persist in many spheres: *Boyle-Mariottesches Gesetz* = *Boyle's Law* = F *La loi de Mariotte*. Note that the German version strikes a compromise between two rival claims.

The really difficult problems lie in the less conspicuous realms of value and conceptual systems. Yet it is precisely here that we are most inclined to take our inherited ideas for granted and to assume that they are universal. The following table illustrates the subtle disparities between the German approach to the marketing of wines, with every German term a variation on the main element *Wert* ($\pm$ value):

Biotischer Wert,	*nutritive values*	(**bioti*. *value,* *enjoyment
(Genusswert)		*value*)
Geschmackswert	*palatability*	(**taste value*)
Marktwert	*marketability*	(**market value*)
Verkehrswert	*product quality*	(**traffic/business value*)

Another example is *Verkehr*, which has a range of meanings that includes traffic, commerce, business, trade, communication, intercourse, circulation.

As against these divergences, it must again be stressed that technology today tends to be international. Countries with a highly developed science and technology tend to resemble each other more and more in way of life, standard of living, habits and customs. This technological 'overlap', together with the internationality of science, undoubtedly facilitates scientific and technical translation. The vocabulary elements that can be termed international are a reflection of this general convergence across linguistic and cultural barriers.

CHAPTER 11

Technical language

᭤᭤᭤᭤᭤᭤

'TECHNICAL language' refers in this chapter to a variety of the general language that has certain definable and distinct features of vocabulary and, to a lesser extent, of grammar. It can further be classified into a number of sub-varieties, according to subject fields, and into three main groups: scientific language; workshop language; and consumer or sales language.[1]

'Language' in this sense does not mean a system on a par with German or English. A technical language is a variety of a general language such as German or English, so we have German technical language, German scientific language (*Fachsprache*) workshop language (*Werkstättensprache*) and sales language (*Verkäufersprache*) and parallel languages in English, French and so on.

Technical languages are sometimes described as occupational dialects. Some technical languages can perhaps be considered as such: the language of fishermen, carpenters, miners, for instance. But technical vocabulary *per se* is much larger than any of the special technical vocabularies of which it is composed, and it is constantly increased by the creation of new expressions. It is probably the most important source of new words in the world's languages today. Yet a dialect in the strict sense has a much more extensive range of expression than a technical language; a dialect is a language in a real sense, a language that has not been standardized and has not been accepted as the official tongue of a state or community. A 'technical language', on the other hand, does not have the range of expression of a dialect; it is a specialized and restricted aspect of language. It cannot be used for writing poetry or even for ordinary conversation, as a dialect, or even a 'dead language' such as Latin, can. The main distinguishing feature of a technical language is its vocabulary, but it also has certain grammatical features that are peculiar to itself or are more pronounced than in ordinary speech.

A technical language might be compared with a style, except that several styles are possible within the same technical language. Scientific language is identified by its formal style, but workshop language tends to have a more casual style, closer to that of everyday speech. Other types of technical language, such as instruction sheets or manuals, advertisements or patents, display variations on the scientific style. Scientific papers are characterized by a rigorously defined use of words and a high frequency of passive forms. Instruction sheets may not use such sharp definitions and the predominant verbal form will be the imperative.

Regional dialects play a part in technical texts, and contribute to vocabularies. German and Italian texts, for example, may contain regionalisms, though dialect expressions are most likely to be found at the workshop or sales level. Older technologies such as watchmaking teem with variations: a *Zeigerstelltrieb* (hand set drive) becomes *pignon coulant* (sliding pinion) in French, *castle wheel* in Britain and *clutch wheel* in the United States. The watchmaker expresses himself differently according to whether he works in Geneva, the Joux Valley, the Neuchâtel Jura or Franche-Comté. The language of a watchmaker in the Black Forest differs quite considerably from that of his counterpart in the German-speaking parts of Switzerland.[2]

A similar observation is made by Savory:

Even a comparatively simple machine may well be composed of dozens or scores of separate parts, and even in different districts of the same country these may not (in fact almost certainly will not) be known by the same name. Hence the description of a vacuum pump translated into Italian may be appreciated in Brindisi and may be far less intelligible in Genoa. The British Standards Bureau makes efforts to reduce this kind of ambiguity. In 1955 one of its recommendations advised a reduction of six thousand names of varied occupations in British coalmining to three hundred, a fact which shows that translation into English of a foreign text dealing with these duties would have nineteen chances in twenty of being misunderstood each time that one of them was mentioned.[3]

The three kinds of technical language – scientific, workshop and sales – do not correspond to social classes. The expressions prevalent in any one of these varieties may be used in the others. The relationship between them is shown in the following diagram:

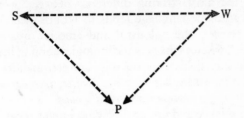

technical language
S = scientific (laboratory) language
W = workshop language
P = sales language

Scientific language is used in research papers and in the exposition of hypotheses and theories. It is normally very formal in style and its vocabulary is highly standardized. But there is a considerable range within this variety and it can be of a high literary standard. Scientific vocabulary includes rigorously defined words and words not usually found in the everyday word stock. It avoids emotional associations and seeks transparency. (Transparency implies that the structure of the expression conveys the meaning virtually at sight.) It makes frequent use of affixes and of words derived from Latin and Greek; this helps to keep the expressions remote from everyday associations and at the same time contributes to internationalism in the vocabulary of science. Yet the bulk of the words originate in the mother tongue, despite the popular impression that scientific language is full of foreign speech elements.

Workshop language comes midway between scientific and general language. It is closer to ordinary speech than scientific language. Scientific discourse tends towards formality and abstractness, whereas both workshop and sales languages, especially the latter, have a colloquial, even sometimes a racy, air. Compared to the precise and cold definitions of scientific communication, that of the workshop is full of spontaneous coinages and metaphor. Metaphor is found in scientific language too, and may play an important part in the formation of concepts. The idea of electric 'currents', for example, is metaphorical and was developed in opposition to the older 'two fluid' theory. But the metaphor in science is rarely recognized as such, whereas it is clearly visible in workshop language.

In fact, the most striking difference between workshop language and scientific language tends to reduce metaphors to precision, to remove the colourful and emotive qualities, to 'demetaphorize', whereas the new technologies abound in neologisms, especially in such fields as aeronautics and astronautics, electronics and nucleonics. Examples here are *booster, crossover network, dead time, breeder, virgin neutron* and *burial ground.* Many of these have passed into science and are probably no longer regarded as metaphor. Workshop level is not a class term; it does not signify a lower status or lower order of intellectual effort. It is a functional term. Workshop coinages are often created by research scientists. There is a degree of correspondence between the workshop level and the more practically minded worker, research technologist, engineer or technician. The workshop level is more applicable to them than to the pure scientist.

There is a partial correlation between scientific language and the research carried out to advance knowledge rather than industry, between workshop language and the research carried out in order to apply existing knowledge to solve an immediate, practical problem. The two are interrelated and applied science is also related to sales and marketing. The development of the microscope, for instance, was a piece of applied science, but it had a profound influence on the 'pure' science of later times.

Specialization in the technical languages is a continuing process. New disciplines are coming into being all the time and at an increasing rate. These specialized languages (or rather vocabularies) interact with one another; terms from one are adapted for use in another; some new specializations are created through the fusion of separate fields (cybernetics, biophysics, medical engineering), though the fields are largely sealed off from one another. Mastery of one subject field does not imply an understanding of others, even if they are fairly closely related. The specialist learns only his own terminology and is a layman as far as other subject fields are concerned. This affects all levels of language, but the consequences are most drastic in science, where the conceptual framework is vital for understanding the text, and the conceptual framework and language are intimately connected. On the workshop level, we can refer to the concrete materials, processes and so on, involved in illumination, but for a text on quantum physics we need to understand the ideas. Understanding the language and understanding the ideas involve virtually the same process.

1 it is specialized and tends to become more and more specialized in contrast to the versatility of ordinary language. Language tends towards liveliness and multiplicity of meaning, but the controlled language of science is manipulated in the direction of insipidity and colourlessness
2 it seeks the most economic use of linguistic means to achieve standardization of terms and usage
3 it seeks to avoid ordinary language associations and endeavours to define its terms accurately

These points find expression in the prevalence of nominalized forms in the syntax and the impersonal modes of expression. Abstruse constructions are avoided and terse sentences preferred. There is also a preference for verbs whose meaning has become so diluted that they are in effect functional rather than lexical (*erfolgen*, *durchführen* and so on). The nominalized style is easier to write and its impersonality avoids commitment to tense, unlike the conversational style. The aims desired in scientific language are not always achieved in practice. Apart from poor style, the workshop and sales languages inevitably introduce uncontrolled usages. They have also inherited all the disorderly terminology and vocabulary of the past. Complete standardization is not without its disadvantages, however. The illogical side of the language is responsible for its creativity. We might have a neat and tidy code, but we would not be able to extend it if language was all logic. Many technical terms, as exemplified in the neologisms above, are spontaneous growths. Some are formed after great deliberation and forethought, as with Faraday's terminology for electricity. But the inspiration of the moment may equally be responsible, and similarly a human desire to be commemorated or to commemorate someone may underlie a scientific word (*hertz*, *ampere*, *farad*). The history of names of innovations in the field of aeroplane, the railway engine, the motor car and so on, provides a fund of illustrations.

Scientific language thus tends to be more logical and lucid than everyday language, but it is not all that logical, and certainly not always logical. The scientist avoids ordinary words and prefers his own, which have acquired exact meanings. He also avoids irrelevant associations. Ideally, scientific words are transparent in both form and structure. They are constructed on agreed principles, built up logically from simple elements, usually Greek or Latin; and the general meaning is understandable from the parts. *Zoology*, for instance, is compounded, like all *-ology* words, from Greek *logo-* = *study* and from Greek *zoö* = *animal*, so *zoology* = animal study. *Biology* includes *bio* = *life*, and means the study of life. These terms are international: F *zoologie*, I *zoologia*, S *zoología*, D *Zoologie* (but also *Tierkunde*). A limited list of Greek and Latin forms will therefore give the scientist access to a huge vocabulary, even if he has had no training in the classics. This applies more to scientific than to workshop language, and more to biology than to nucleonics. It is only partly true for some disciplines: for example, in chemical nomenclature:

Formula	D	E	F
HCOOH	*Ameisensäure* (*Ameise, ant*)	*formic acid* (L *formica, ant*)	*acide formique*
CH₃COOH	*Essigsäure* (*Essig, vinegar*)	*acetic acid* (L *acetum, vinegar*)	*acide acétique*
H₂SO₄	*Schwefelsäure* (*Schwefel, sulphur/brimstone*)	*sulphuric acid* (L *sulfur, brimstone/ sulphur*	*acide sulfurique*

This table offers an example of synonymy across languages. The same phenomenon occurs in technical language, which is not free from polysemy and homonymy. Ambiguous terms are particularly common in the biological sciences. Let us take the case of *il porcospino* in Italian, which is by no means rare. It may mean the same as either *il riccio* (*Erinaceus Europeus*) = *hedgehog*, or *l'istrice* (*Hystrix Cristatus*) = *porcupine*. Polysemy may also be inter-disciplinary, as we have seen with *plasma*. Another example is *hysteresis*, which is used in both electrical engineering and colloid chemistry.

The sources of vocabulary in technical language are: nominal phrases; compounds; derivatives; new applications of words (*force, work, current*); neologisms (*kodak, nylon*); borrowings (from classical and modern languages). A borrowing may be a direct transfer (English *science*), or it may be a loan translation (German *Wissenschaft*). A word borrowed in the first way will quite probably change its meaning in the new language. Thus the meaning of Latin *scientia* (which originally referred to knowledge in general) became confined to knowledge acquired by the methods used in the natural sciences. The first three procedures may use elements from the last. This will indeed be the tendency in the scientific language, but in the workshop and sales languages native elements will be used.

The difference between scientific and workshop terms is seen in the expression *spark plug*: German *Zündkerze* ('ignition candle'), French *bougie d'allumage* and Italian *candele d'accensione*. The determinant element ('ignition') refers to the purpose and the main element ('candle') emphasizes the function. 'Ignition' is omitted in English because it is a scientific term, a laboratory term. Instead the easily grasped workshop term *spark plug* is used. This does not lay the stress on the function but on its appearance (plug and spark). 'Ignition' is more abstract than 'spark'. The continental expressions are nearer to the scientific language than the English.

The scientific language draws on a humanistic education, while workshop terms are non-literary, practical, colloquial and sometimes humorous. They are the heirs of the old artisan languages and many of the forms created by the old craftsmen survive in the workshop vocabulary. Milling engineering is an activity, a *Gewerbe*, that falls into this category, and it is full of picturesque words and phrases:

Einlaufschlauch ('inlet tube')	sleeve (British), stocking (USA)
Nachmehl ('after flour')	low-grade flour, superfine wastings (British), Red Dog (USA)

Milling vocabulary contains terms like *damsel* (device agitating the hopper); *eye* (central opening of millstone); *hopper boy*; *bosom* (area immediately surrounding the eye); *concert chest*; *shoe, skirt*. Today milling is an industry using the most sophisticated techniques and it has a terminology that conforms with this in addition to the above.

Homely terms are still found throughout the engineering industries:

Absetzer	*spreader*
(*absetzen* = deposit)	
Brecher	*crusher*
Greifer	*grab*
Feinhubgeschwindigkeit	*creeping speed*
('delicate/fine lifting speed')	

The last item again shows a contrast between levels. English is far more down to earth than German. German provides a more accurate definition of the process, but in general this is not necessary in workshop language, where the aim is to give a catchy label. *Brecher* and *Greifer* are on the same conceptual level as *crusher* and *grab*.

A science or technology will not consist exclusively of words on one level. The new disciplines in particular are very catholic in this respect, no doubt because of the tremendous number of new words they use and the speed with which they are produced. Astronautics, for instance, has words of Latin and Greek origin, like *astronautics* itself (*Astronautik*), *extragalaktisch* (*extragalactic*); words from earlier vocabularies, modified for new use, like *Exzentrizität* (*eccentricity*), *Orbit* and *Konjunktion* (*conjunction*); coinages from the mother tongue, like *Erdschein* (*earthshine*), *grayout*, *Mondfinsternis*; translations like *burnout* from *Brennschluss*; borrowings from one language to another, like *cosmonaut* and *sputnik* from Russian and *astronautics* from French *astronautique*; names of people, like *Compton effect* and *Frauenhofer-Linie*. Many of these terms, especially those from modern languages, are of the workshop variety.

Sales language is aimed at persuading the reader to purchase a product or service. The purchaser may be a technical expert, but he may equally be an administrator, manager or executive with minimal technical knowledge, to whom scientific language would be largely unintelligible. Scientific terms are not used merely to convey information. They are often used to impress. Sales language is more individualistic than the other two varieties, so its coinages are even more colourful than those of the workshop. Many are ephemeral and will disappear once they have served their purpose.

Nevertheless many words in the sales language have passed into the general language, or even into the technical language. In some instances a deliberate attempt was made to provide a lasting term, but in any case some appellations have endured, while others have not. *Nylon* has entered into the general vocabulary, but *kodak* has not become a term for cameras in general. Some terms are built up from recognizable elements: *Zelluloid, celluloid* (cellulose nitrate and camphor); *Buna* (butadiene and Natrium = sodium). There is an enormous demand for new terms in the field of synthetic substances. *Sicherheitsglas* (safety glass), which has become a generally accepted term, is much more striking than, say, *laminated glass*; and it illustrates the point that in the sales sphere, the stress is more on result and function than on structure and process of manufacture. (This is not true of *celluloid* and *Buna*, of course, but these are merely convenient labels and easy to pronounce.) The *safety* in *safety glass* naturally has great publicity value because of the emotive colouring of the determinant element. For the same reason, elements like *echt* (genuine), *fest* (*tight*), and *frei* are popular in word formations here: *wasserfest* (*watertight*), *farbecht* (*colour fast, unfading*), *rostfrei* (*rust-free*).

Sales language is characterized by its syntax. It uses an inordinate number of verbless (exclamatory) constructions:

> *components for belt conveyors*
> *Bühler Transport- und Entladeanlage*
> *anschlussfertige Kranführerkanzel für die elektrische Ausrüstung von Krananlagen*

Statements resembling the informative sentences of workshop prose are, however, also common:

1a *Alle unsere Schrapper der gleichen Type in Zwei- oder Dreitrommelausführung besitzen die gleichen Bauelemente, so dass die Teile untereinander austauschbar sind*

1b *All our two and three drum scrapers of identical design incorporate interchangeable construction elements for the economic solution of your spare parts problem*

Slogans without a finite verb but with participle construction are again common:

2a *Gross-Schaufelradbagger für eine Tagesförderleistung von 110,000 t Braunkohle in der Montage*

2b *Giant bucket-wheel excavator designed to cut 110,000 tons of lignite per day, in the course of assembly*

It must be remembered that sales language also tends to rely to a great extent on illustrations (drawings, photographs and diagrams), and is by its very nature dramatic. The emotive element is less prominent when it is designed to appeal to people with technical training but even with them, publicity tricks used on the general public have been found to be effective.

Associated with the use of trademarks is that of alphabetical letters. This is in keeping with the trend towards economy in technical language, which also finds expression in abbreviations and acronyms.

The translation of sales language becomes more difficult the further it departs from the more or less sober style of the technical writer seeking to inform his peers. It has a much more national flavour than these, using idiom and metaphor freely in a personal approach. When it is aimed at engineers or executives of industrial firms, the emphasis must be on the less romantic attractions of economy, reliability and flexibility. Drama is contained in the imperative style; the exclamation (sometimes simply the firm's name; perhaps a brief statement on performance (*Dauerleistung 75 to† stündlich = continuous output 75 tonnes hourly*); a picture of the equipment in action; and possibly layout. Presentation for a wider circle of readers also depends to a greater extent on plays on words, on associations, on national customs. One German beer advertisement starts with the words *Bitte ein Bit!* (**Please a bit!*), an untranslatable play on words that is just as embedded in the language as *Don't be vague – Ask for Haig*. The use of loan words makes translation even more difficult because they have a value in the one language that they cannot possibly have in the other. *Start Nach Shanghai + Bangkok* is racier than *Take off for Shanghai + Bangkok*, because *start* has an exotic tang in German that it does not have in the original English and that *take off* does not possess. Similarly, *das graue Alltag* is not quite the same as *the grey everyday* or *everyday greyness* or *the grey hue of everyday*, and we cannot say *sonnenblauer Sommer* in English, because *sun-blue summer* does not have the same effect.

The translator's task in such cases is not to achieve a one-to-one, two-to-one or one-to-two correspondence, but to absorb the mood of the utterance and to attempt to recreate that mood in the TL.

† Usually written t.

TECHNICAL SYNTAX

The technical style as practised is horrifying from the aesthetic point of view. It is the exact opposite of literary style. Its aim is to create an average, mediocre, impersonal style, with the individual kept as much in the background as possible. It should be made clear at the outset that there is good and bad technical style, but merit lies entirely in fitness for purpose; thus a style that is bad from other points of view is good if it communicates exactly what the author wants. For instance, a certain woolliness in patent documents might be exactly right, as the author's aim is to obscure rather than to clarify. But on the whole good technical style is clear.

Technical prose tends to be written in terse sentences. Two-clause sentences are popular. Condensation is largely realized in German by morphological means, as in *Endbearbeitungskosten = finishing cost*. Looser combinations are used in English, as in *alternating current motor design*.

The role of the verb is reduced in most sentences and a nominalized sentence pattern is characteristic. Types of nominalized expression include:

Action noun replaces verb

Instead of V, N + V. The verb is used like a functional verb and has virtually only a linking effect. Verbs with a very general rather than a specific meaning are common: *bilden (form)*, *darstellen (present)*, *dienen (serve)*, *entsprechen (correspond to)*, *bestehen (consist of)*. These may represent up to three-quarters of the verbs in a text. The noun is often of the *-ung* type, or a nominalized adjective expressing an abstract idea in *-heit/-keit*:

Er führt eine Untersuchung durch . . .
He carries out an investigation into . . .

Verbal predication changed

Infinitive or participle replaces sentence:

Um Reizwirkungen zu vermeiden, zieht man Zinkpaste vor
To avoid irritation, zinc paste is preferred

Noun replaces sentence:

Zur Vermeidung von Reizwirkungen zieht man Zinkpaste vor
For the avoidance of irritations, zinc paste is preferred

Apposition

Eine relative einfache Verallgemeinerung dieser bisherigen Quanten-
mechanik, die Paulische Theorie des Elektronenspins ...

... a relatively simple generalization of our quantum mechanics, the
Pauli theory of electronic spin

Nominal instead of verbal clauses

Verb completely superseded. Examples are entries in handbooks, captions to illustrations, advertisements, lists, tables, biological diagnoses (which are still in Latin), workshop reports:

Äthiopien: Staat in Nordostafrika, konstitutionelle Monarchie ...
Ethiopia: State in North East Africa, constitutional monarchy ...

We have already seen that the passive is preferred. This is in keeping with the tendency to formulate the sentence in terms of theme and related event, rather than in terms of agent and action, which is the normal pattern in ordinary utterance. A state of affairs can be emphasized without the author mentioning the agent. This is also in keeping with a desire for economy, which is most strongly manifested in verbless statements. The agent can be weakened in German by means of sentences with *man*, and *es + sein* constructions, in English by means of the passive *there + be*.

The present indicative S+P+Comp sentence is well represented in this type of language, and so are *dass-* (*that*) clauses, relative clauses and conditional clauses. Prepositional groups play an increasing part both in German and in English:

um Stahl zu gewinnen	*zur Stahlgewinnung*
to reclaim steel	*for steel reclamation*
wenn man Wasser auf 0° Celsius abkühlt	*bei Abkühlung des Wassers auf 0° Celsius*
when water is cooled to 0° Celsius	*with cooling of water to 0° Celsius*

The nominal style is more abstract than the verbal and easier to write. There is more diversity in the verbal style, but while this is excellent for self-expression, the technical writer prefers to avoid the intrusion of the individual. Nominalization helps to make the writing seem something special and esoteric. In past centuries, Latin was the language of learned communication and was used to provide the specialized means deemed necessary. Today this role has been taken over by the nominal style.

CHAPTER 12

Standardized terminology

๑๑๑๑๑๑

THE aim in technical language is to achieve the highest degree of precision in the use of words. A scientist will often express his irritation and impatience with the vagaries of the language he has to use. He feels that ordinary language is a very inadequate, inefficient and frustating tool. He feels a strong urge to improve this tool, to turn it into a more effective instrument, to make it do its job properly.

The deliberate standardization of terminology arises out of this need. It is an attempt to clear the channels. Even ordinary language implies a tendency to standardization. If there was no stability in the units and their rules of combination, we should not be able to talk to one another and understand one another. The relation of unit and meaning must not vary too much, or such communication will become impossible. The very concept of a language system implies a degree of standardization. The rate of change must be slow. But if there were no change at all, if there were too high a degree of standardization in the ordinary language, it would lose flexibility, and become cumbersome and ineffectual. The language system is therefore a compromise between the demands of standardization and fixation on the one hand, and a tendency to innovation and change on the other. The arbitrary element in the meaning of words has advantages. A word does not inherently enshrine a specific meaning. In principle it can mean anything society chooses it to mean. Yet it must not be so arbitrary that we do not know what it means at any given time. There must be agreement that over a certain period, at least, say a generation, a word will restrict its meaning and not extend it too far beyond a certain range.

Standardization is strongest in specialized languages, where precision and rigidity of meaning are valuable and this has produced a strong incentive to standardize. It is weakest in areas such as advertising, journalism or slang, where fashion and the desire for novelty prevail. The degree of standardization demanded of technical languages today is far greater than in the general language. It is far greater in fact than in specialized languages in the past, when the standardization processes were left to the interplay of blind forces and individual efforts. This is a direct outcome of technology and science in language. The immense proliferation of technical words necessitates deliberate attempts at standardization based on a proper study of the issues. Communication failure is an ever-present risk today. Obstruction, not only of the channels between science and the public, not only between specialization and specialization, but within a single specialization, is a common occurrence.

One writer has estimated that electrical engineering alone uses more than fifty thousand technical terms. 'Each of these represents a different concept and no more than one hundred of them can have existed when Faraday conducted his famous experiments. A substantial proportion of them are less than ten years old.'[1]

But quantity is not the only factor. As we have seen, ordinary language carries within it a host of assumptions, predispositions and prejudices. It is good for colloquial language to have a certain vagueness. Vagueness makes for flexibility and economy of effort. We would experience great tension if we had to be precise in every utterance we made. In certain everyday situations precision is required and little imagination is needed to conceive what life would be like if we had to be on our guard in this way all the time. On the other hand imprecision and slackness represent a hazard in the technical sphere, where we have to remove ambiguity and vagueness from words as far as possible and use them like instruments.

Standardization of terminology is an urgent need, both in old fields, where the language has grown up in an uncontrolled manner, and in new disciplines, where words spring up daily like mushrooms (toadstools would perhaps be nearer the mark for some coinages). New fields are constantly being opened and the existing ones become highly complicated, resulting in growing communication difficulty. Moreover rapid development is occurring all over the world. Standardization is an international problem.

These problems are very important from the economic point of view, for it is essential that intelligible and unambiguous language be used for framing commercial specifications and contracts. This naturally applies with equal force to the rendering of these documents from one language into another.

A technical term is a special type of term. It has some of the characteristics of a name, and is a kind of label. It is also sometimes an index to a complicated idea. *Feedback* is a handy label standing for a circumlocution: 'For any system which converts energy and has an input and an output, the return of a fraction of the output to the input.'[2] But a term is more than a mere label. *Feedback* is not an empty term for which a number of others might be substituted; it also contains some of the idea for which it stands. Expressions that have been standardized and have become mere labels may be called terms if they are of the type used in chemistry and biology and then belong to an organized system of nomenclature. A nomenclature system differs from a terminological system on the lines suggested above.

The term always stands for a concept, and it should preferably stand for a single concept. This concept has a meaning that is fixed, abstract and general. Only features that are considered to be relevant are retained in the meaning; other features are jettisoned. The meaning sought is denotational meaning, so all other kinds of meaning are suppressed. When a word with designative meaning is transformed into a term, the idea is to get rid of that meaning and to leave only the reference to the concept. The aesthetic and the emotive elements are a hindrance in the creation of terms.

The guiding principle of the term is that it should be accurate and convenient. Service prose can be evaluated in terms of engineering criteria, in terms of performance and energy. These are the criteria used by people such as Dr Eugen Wüster, who has probably done more to give terminological standardization a scientific basis than anyone else in our day, having the advantage of training both as an engineer and as a linguistic scientist. Dr Wüster calls a slight expenditure of energy in the communication process *Bequemlichkeit* (convenience). A slight loss in the content (load) is *Genauigkeit* (accuracy).[3] These terms are of the same kind as those used in the standardization of products and processes in general. The accuracy of a message is proportionate to the correspondence between the ideas intended by the sender and the ideas understood by the receiver. Any link in the communication chain has a degree of convenience and accuracy. In an ideal situation every link would be convenient and accurate to the utmost degree. This is of course Utopian. In practice, convenience and accuracy are almost always conflicting demands, and the best that can be achieved is a compromise between the two. In everyday language we require the maximum convenience, even at the cost of accuracy. Everyday language therefore tends to be vague but easy to manipulate, while scientific language sets a premium on accuracy and is therefore difficult to manipulate. This is less of a handicap in written communications that it would be in speech. Diagrams used for chemical structures cannot be rendered in speech, and mathematical equations, too, are visual devices.

A language system that opts for greater accuracy at the expense of convenience is uneconomic or inefficient. A language in which great accuracy cannot be achieved is inadequate for the needs of science. The best system is one that allows a choice in the degree of accuracy best suited to the particular circumstances. Requirements vary for scientific, workshop and consumer languages. Clearly scientific language prefers accuracy, while workshop language will sacrifice some accuracy to achieve convenience and consumer or sales language may ignore accuracy altogether for the sake of convenience.

In technical language, term and concept are inextricably associated. Ideally, term construction follows the pattern of concept construction in the sciences. The terminology of a science is part of its method, of its processes of discovery. It is an essential element in the conceptual framework of the science. Faraday created a new terminology for electricity because the old terminology implied ideas that he had disproved. In workshop and sales languages this aspect is not so important, since in certain circumstances accuracy may be not only an expensive luxury but an inhibiting factor.

Scientific terminology is prescriptive. Ordinary language dictionaries are said to be descriptive, not prescriptive: they describe how the language *is* used, not how it *should be* used. Modern grammars also purport to describe usage, not to lay down rules. Technical terminology, on the other hand, makes no bones about being prescriptive. The only qualification is that some terminology has institutional backing, and therefore virtually has the force of legislation, while other terminology does not. Many lexical items used in technical language have never been standardized or properly defined. Where rules are laid down, these are very strict and the author or translator departs from them only at his peril. The terminologist of science does certainly record usage, but also, wherever this is possible or necessary, he attempts to create it.

The scientific term therefore has a unique role. It is a member of a conceptual structure and an integral part of the discipline. A term is defined by the content it stands for and not by any peculiarities in its linguistic structure. A good example given by one writer on the subject is that of sulphuric acid and orange juice: both these substances are chemical compounds; both are subject to analysis and reduction to simpler constituents; but a statement that sulphuric acid is a chemical compound would be regarded by a chemist as true, whereas a statement that orange juice is a chemical compound would be regarded by a chemist as an inconsistent combination of a non-term with a term. The two expressions refer to individual, clearly recognizable substances: 'But the science of chemistry does not frame its theories in terms of orange juice as an individual substance, and accordingly its name has no status as a chemical term. Sulphuric acid, on the other hand, is a substance which has been singled out along with many others as significant and useful in the making of a chemical theory.'[4] An expression is a term only within the framework of a conceptual system in which it is defined as a term. In the framework of chemistry, sulphuric acid is a term, but orange juice is not. Yet within another framework, say, food technology, both sulphuric acid and orange juice may have the status of terms. Membership of a terminological system thus gives an expression terminological status. Without it, it is a lexical item, not a term. A term is always a lexical item as well, and can be treated as such within sentences and phrases. Terms also function as word classes (usually as nouns but also as verbs or other parts of speech), and they can operate in syntactic relations (as subject, object and so on of a sentence). But a word becomes a term only when it is properly defined as a member of a terminological system.

This system has approximately the same meaning for a term as context and situation have for a lexical item; but this does not imply that a technical term does not function as a lexical item in the ways described above at the same time as it functions as a term. Hence context and situation may in certain cases play a role in identifying terms. It is important to bear in mind that numerous terminological systems exist and that a term may not have the same meaning in one as in another. For example, *oilseed plant* has one meaning in the terminological system of botany or agriculture and another in that of chemical engineering.

Definition plays a role of utmost importance for the term. This point will be better illustrated by Bloomfield's statement: 'A technical term . . . replaces long phrases, or even a complicated discourse, and its meaning is fixed by an agreement of definition, which, in science, receives explicit formulation and strict adherence.'[5] This brevity is especially significant for the term. In this respect workshop language often goes even further than scientific language and displays even greater brevity.

A concept is first of all defined, and subsequently given a name. That is why translation of scientific terms, and indeed scientific expression in general, involves transferring concepts, not merely words. To define a concept we must demarcate its boundaries in contrast to allied concepts and distinguish it from these. The progress of science has always been associated with the development of new concepts. 'Each concept has been carefully isolated from neighbouring ones; the concept of acceleration from those of movement and velocity. Each of these has been given harder, clearer outlines than any concept in humanistic studies could ever achieve. Each has been defined in such a way as to preserve its uniqueness.'[6]

This rigour in the definition of concepts helps to make translation more feasible. Another line of development in the sciences, a tendency to greater and greater abstraction, which is particularly marked in the older sciences, makes it more difficult. In physics the main concepts have been reformulated and unification has been achieved in the theoretical framework. But this has led to more abstract language, which is much more difficult to grasp than that of classical physics: 'To-day modern physics has completely passed from the realm of ordinary language into a mathematical symbolism often incomprehensible even to the physicist who is not an expert in the field the particular theory describes.'[7] Mathematical equations as such do not present a problem for the translator, since they are not translated but represent a universal symbolism, but the accompanying highly abstract prose is always difficult to translate.

In general a term must lack ambiguity, i.e. there should be only one concept to one term. Although a term is not necessarily a word, but can, linguistically speaking, be a syntagma, it must be meaningful, simple and euphonious. It should also preferably be international and capable of forming derivations.

Examples of ambiguity are older terms in physics such as *Kraft, force*; *Arbeit, work*; and *Ladung, charge*, which have been described as 'incitements to inexact thinking'.[8] Others become more and more troublesome because of their provincialism. The use of such 'an anthropomorphic name' as *work* for a scientific concept has been called highly reprehensible. 'The concept is surely a special case of energy, and it can always be called by that name.'[9]

When a term leaves its terminological system, it loses its character as a term. If it enters another terminological system, it acquires a new denotation. The transition of terms from system to system is a constant process. Terms pass into workshop language, sales language and the general language all the time.

Many terms are formed by abstraction from ordinary language expressions. Certain possibilities within the range of meaning of these items are suppressed and the expression is limited to only one meaning. This meaning is clearly defined, much more clearly than in the ordinary language. This point is illustrated by the difference between the way in which we define dimensions in everyday life and the technical explanation. *Hill* and *mountain* are relative terms and are highly subjective. It will not be accurate to translate Italian *monte* as *mountain* and *colle* as *hill* unless we know what these words represent to the writer, and what other terms for projecting land masses exist in the vocabulary. As a geographical term, however, a hill is anything rising less than 600 metres above level, and anything above that height is a mountain. Similarly in mechanics *work* is the product of a force and the distance through which it moves. It is distinguished from *energy* (capacity for doing work) and *power* (rate of doing work). One property is selected from the bundle of ideas that surround the word in the general language and is developed along certain lines.

We have noted transferred meanings. *Crane* and *nut* once referred to a bird and to a botanical product. The device *crane* received its name because of a physical resemblance to the bird, but in due course, as cranes became more familiar as devices, this became the dominant meaning. Few people today would associate the two meanings at all. The same applies to *nut*.

Metaphor is a time-honoured method of creating terms, perhaps the earliest of all. Workshop and consumer language, as we have seen, constantly resort to metaphor, but scientific language uses more metaphor than appears at first sight. Faraday's *field* was just as much a metaphor as the *fluid* that it displaced. Metaphors are indispensable to technical language. They tend to correspond in scientific language (*Wellenpakete, wave packet*), but there may be one-to-nought and nought-to-one correspondences. *Die Wellengruppe zerfliesst* is more abstract (*Wellengruppe, wave group*) and more metaphorical (*zerfliesst, melts, disperses*) than *the wave packet spreads out*. Compare *das Teilchenbild* (the particle picture) with *the particle theory*. The disparities in workshop and sales language are generally even more marked.

The process of standardization tends to destroy the vividness and pictorial qualities of the metaphor, transforming it into an expression of conceptual exactness. The metaphor gradually fades away, as with *crane*, in its industrial employment. The names of many chemical compounds and even elements are examples of dead metaphors. Who now spontaneously associates *formic acid* with ants?

When a term is used loosely in common parlance, it acquires a whole range of meanings in addition to its scientific meaning. These usually drown the strictly defined one and serve little purpose except to impress. That is, their conceptually clarifying function is lost and they acquire an emotive function instead. This may be deliberate, as when scientific expressions are used in advertising to give an aura of authority to some claim, or it may be inadvertent, as when the user has picked up the term somewhere and is ignorant of its proper use. Journalists are great offenders in this respect, but then so are technical people who use their special language in the sphere of ordinary communication.

The expression *concept* – a term used in philosophy and psychology – has become popular in advertising and workshop language. 'A new concept in furnishing' accompanies 'a new concept in computer design'. This is simply a more pompous way of saying 'a bright new idea'. The very word *philosophy* has been wrested from its moorings and drawn into workshop language as a substitute for 'principles', 'mode of operation' and so on ('the philosophy of this equipment'). Unlike authentic terminology, this is not a contribution to language efficiency, since it introduces unnecessary confusion into general language. The word *philosophy* becomes ambiguous and vulgarized. It becomes difficult to use it for the purpose for which it has been used for centuries. Eventually, having left this trail of confusion behind it, it goes out of fashion. It may then revert to its original use. It is pointless to complain. The translator must deal with the facts and know that *Grundprinzipien*, or *Methoden*, might in some contexts have to be rendered as *philosophy*.

Transistor is another illuminating example. In the vocabulary of electronics it denotes a semi-conductor with specific properties, but in everyday language it has come to mean a radio receiver. It is irrelevant whether this receiver contains transistors or not. The layman will not even know what a transistor looks like, or rather he will assume that it is merely a portable radio set.

When the converse process occurs an expression is incorporated into a terminological system – receiving a clear definition – and becomes a term. *Energy* is a word in ordinary language but a term in physics.

In the general language the word *worm*, as we have seen, can mean many things. *Webster's Seventh New Collegiate Dictionary* offers for the noun *worm* the following entries: an annelid worm; several other types of creatures; a human being who is an object of contempt; something that inwardly torments; snake; helminthiasis; something spiral or vermiculate in form; a type of revolving screw; a spiral condensing tube used in drilling; Archimedes' Screw and so on. We may be able to tell from the context which *worm* is meant, but we may need extra-linguistic knowledge – the total communication chain. The terminological system can be substituted for the context and situation. It represents in effect the subject field as a whole. But a technical term does not mean the same thing in all terminologies. *Worm* does not mean the same in engineering as in biology or in medical science.

The definition process selects only the characteristics relevant to the specific field and eliminates everything else. The polysemy that exists in the lexical term disappears and the expression relates to only one concept. This definition of the concept, as we have seen, is the prerequisite for the naming of the term.

There is a resemblance between terminological systems and the structures and fields found in the general vocabulary. But the items in a terminological system exist in a strictly logical relation to each other and the system. An ideal terminological system would display this logical construction, and all terminological systems approximate in varying degrees to this ideal. In lexical sub-systems the principle of association is only partly logical. It is based on incidental features, superficial characteristics: the whale swims, so it is a fish; the sun, moon, Venus and Mars are in the sky, so they are stars. It is also based on emotional associations and other considerations that play no part in forming a terminology. A terminological system may be part of a more comprehensive system, even as some disciplines have sub-disciplines and may be part of a larger body of learning. Nuclear physics is part of physics, electrical engineering is a branch of engineering, embryology is a sub-division of zoology.

Since the system has the effect of an extra-linguistic situation, a properly compiled terminological dictionary can provide more reliable translation equivalents than an ordinary dictionary.

Terminologies are associated with conceptual systems, whereas nomenclature consists of the labels given to various objects. Nomenclature forms a system in the same way as terminology does for a particular field. The desiderata for terminology and nomenclature overlap to a large extent: examples here are internationality and lack of ambiguity. Nomenclature also uses Latin and Greek formative elements to create expressions. Nomenclatures used in zoology, botany and chemistry show this clearly. These systems are based on agreement and are no longer associated with the common language. Botanists still employ Linnaean principles for designating discoveries and describing them.

Nomenclatures operate in the same way as terminologies in general. Removed from the system and introduced into a word field, a term becomes a word (or phrase), as 'oxygen' does in the following sentence: 'It's good to get oxygen into one's lungs again.' This is scientifically absurd. If an item from nomenclature enters a terminological system, it becomes a term.

The relationship between nomenclature, terminology and words is exemplified in *macadamize*, named after John Loudon McAdam (1756–1836). *Macadam* (*Schotter*) denotes a process in road-making, *macadamization*. Compare E *macadamize* with D *makadamisieren, beschottern*.

The language of the workshop is less highly terminologized and less controlled than scientific language. There is far more spontaneity and disorder in its expressions. Nevertheless workshop talk plays an important role in language creation.

Workshop usage tends to convert complex standardized terms into more picturesque ones, not least by means of abbreviation. Accurate terms can be clumsy. The language of the workshop is related to conversation and practical concerns, that of the scientist to writing and theories. *Crane, nut, worm* and the like belong to the workshop order of expressions, while *photon* and *probability density* are scientific coinages. The former are not conceived as part of a logically organized system of thought; they arise to meet an immediate need. In our own day many ancient crafts have been transformed into modern industries employing highly trained experts. These experts, unlike the old artisans and craftsmen, have been trained in scientific method and thinking, and yet they have inherited the old terminology. To mould these old expressions into sharply defined terms becomes a major problem. 'Attempts to restrict the meaning of a mining term – as in *ore* – or to extend it – as in *mineral* – are a common source of confusion and inexactitude.'[10]

A glossary in one well-known textbook on milling engineering[11] provides not only equivalents in French, Spanish and German but also American and Canadian terms where they differ from the British ones. A German glossary presents the American term in parentheses alongside the British term:

Abstoss	*overtails (tailings)*
Auszugmehl	*first-grade flour (patent flour)*
Mais	*maize, corn (corn)*
Riffeling	*fluting (corrugation)*
Zyklon	*cyclone dust collector (centrifugal dust collector)*

Fields where terminology is poorly organized inevitably create difficulties in translation. Medical terminology is a case in point. The specialists in this discipline are themselves well aware of the need to bring order out of chaos. The task of standardizing terminologies is of course undertaken by representatives of the disciplines, generally on an international basis. An International group was set up by the Council for International Organizations of Medical Science under the auspices of Unesco and the World Health Organization (WHO), to standardize its terminology.

Many other professional organizations have also concerned

themselves with terminology, including the International Electro-technical Commission and the Conférence Internationale des Poids et des Mesures (CIPM) (concerned with mass, length and electrical units). The standardization of terminology is very much linked up with the standardization of units, and the national standards organizations of the various countries concern themselves with terminology. These include the American Standards Organization (ASA), the British Standards Institution (BSI), and the DNA (Deutscher Normungsausschuss), which also publishes DIN standards in ter-minology and nomenclature. On the international scale a special committee of the International Standards Organization (ISO) is concerned with principles of terminology (ISO/TC37).

Internationalism is a very important aspect of technical lan-guage. Science and technology are international and depend on international communication for their progress. It goes without saying that internationalism in technical language is important in translation and can make the translator's task much easier.

A very large vocabulary in the various languages can be called international. This is so much in evidence that the Third *Webster's* (1961) uses the label ISV for International Scientific Vocabulary: 'A part of the vocabulary of the sciences and other specialized studies that consists of words or other linguistic forms current in two or more languages and differing from New Latin in being adapted to the structure of the individual languages in which they appear.' The words *transistor, eccentricity, orbit* and *astrobiology* all belong to this vocabulary. Parts of some phrases are also inter-national (*Mach number, Mach-Zahl, nombre de Mach*) not to mention suffixes and prefixes. Chemical symbols and formulae are inter-national, as are mathematical equations. There are generally variations in orthography, but expressions can be recognized: I *elio* or *helio* = *helium*; I *ossido* = E *oxide* = D *Oxyd*; D *sulfat* = E *sulphate*; D *Eigenfunktion* = E *eigenfunction*; D *Quantenphysik* = E *quantum physics*; D *Resonanz* = E *resonance*; S *escandio* = E *Scandium*; D *Äthylen* = E *ethylene* = I *etilene* = F *éthylène*.

Partial internationality also occurs in some areas. German, for instance, tends to preserve its own national terms alongside the international alternative. Chemical names are examples:

See table overlea,

Note that the list is not completely consistent. The English word

	D	Ne	Sv	E	F	I	S
C	Kohlenstoff	koolstof	kol	carbon	carbone	carbonio	carbono
H	Wasserstoff	waterstof	väte	hydrogen	hydrogène	idrogeno	hidrógeno
N	Stickstoff	stikstof	kväve	nitrogen	azote	azoto	nitrógeno
K	Kalium	Kalium	Kalium	potassium	potassium	potassio	potasio
Na	Natrium	Natrium	Natrium	sodium	sodium	sodio	sodio
Fe	Eisen	ijzer	järn	iron	fer	ferro	hierro
Pb	Blei	lood	bly	lead	plomb	piombo	plomo

for Fe is similar to the Swedish, and that for Pb is similar to Dutch. In general, however, English conforms to the Romance pattern.

Sometimes loans develop separately in individual languages. International standardization helps to stem this process, but without it variations soon appear, especially on the workshop and general levels. Technical languages must be intelligible to technologists as well as to academics. A term tends to develop an equivalent for the technically uneducated, as with *electronic brain*. The man in the street tends to simplify or shorten long expressions: *Photographie* becomes *photo* or *foto*; *Automobil* becomes *Auto* and so on. Some forms acquire different ranges of meaning, and these can be very deceptive. D *Phosphor* is the same as *phosphorus* in English, but S *fósforo* means *match* as well, while I *fosforo* signifies both *match* and *morning star*.

CHAPTER 13

Translation procedures

☙☙☙☙☙☙

TRANSLATION procedures are the technical devices used to transfer the meaning of a text in one language into a text in another language. They involve essentially adding structural or lexical elements to those present in the SL or subtracting from them; eliminating elements that are obligatory in the SL but unnecessary in the TL or with no counterpart there, and where disparity between the two media goes beyond language patterns, adapting the content of the message so that the TL text will come as close as possible to the intent of the SL text and create a similar impact.

The transfer between two expressions can be expressed grammatically, by lexical means or by direct reference to the extralinguistic situation, or by any combination of these.

Translation in the very broad sense of the term can be graduated in terms of different levels of complexity (see Chapter 4):

Transcription

This denotes rendering the sounds of an SL into a TL form. It mainly concerns names in scripts such as Arabic, Hebrew, Chinese and Japanese, but it may also be necessary to transcribe from a language using, say, a Latin alphabet into another language using a Latin alphabet. Languages using basically the same alphabet may not have the same range of sounds. Sounds from any language rendered in a phonetic alphabet are transcribed into it. When we write *Fujiyama*, we are transcribing from Japanese into English. In German the name appears as *Fudschijama*. Many place names, even in closely related languages, are transcribed forms. Since the phonological system of an SL may be very different from the TL's, the transcription may sound nothing like his own pronunciation of the name to the speaker of the SL. The transcription is made in terms of the TL phonological system. There are certain established conventions for the transcription of certain names into English, but generally it is an approximation to the SL pronunciation. Thus *Braunschweig* becomes *Brunswick, München, Munich,* whereas *Berlin* remains the same.

Variations occur between the languages employing Latin alphabets. The letters are mostly identical, but each language has several letters or signs that do not occur in the others. German has the *umlaut* over vowels a, o and u, as in *Mädchen, grösser* and *füllen;* Swedish has the ä, ö, å.

Transliteration

Transliteration is the process of rendering the letters of one alphabet in the letters of another with a different alphabetical system. No transliteration is needed between languages sharing the same alphabet, such as German and English. But Russian has to be translated into a language using a Latin alphabet, as does Greek or Arabic, Hebrew and Devanagari. These different alphabetical systems form one of the barriers to translation. In our day transliteration is mainly from the Russian Cyrillic alphabet into the Latin alphabets. It is also useful to know the values of the letters of the Greek alphabet, since these are used very extensively in scientific symbolism. Rendering a Russian proper name into English may entail both transliteration and transcription. The English word *Kremlin* is derived from Russian *kreml (citadel),* of Tartar origin.

Borrowing

Many types of borrowing are made from one language to another, a procedure that is often resorted to when the TL has no equivalent for the SL units.

The borrowing of items lock, stock and barrel: English has borrowed *Blitzkrieg, Brehmstrahlung, Brennschluss* from German without change.

It may be of a structural element, like E and D -*er* from Latin -*arius* or D -*ieren* from F -*ier.*

It may be of the concept, as with *set theory* (on the model of German *Mengenlehre*)

It may be conceptual and structural, as in *superman* (from *Übermensch*).

Literal

This is one-to-one structural and conceptual correspondence. It can include borrowings and word-for-word translation. Word-for-word translation takes the word as the unit of translation, on the assumption that there is a corresponding word in the TL for each SL word. This presupposes a kind of interlingual synonymy. It is the popular view of translation. We do get one-to-one correspondences for properly standardized international terms, but technical texts also contain ordinary language elements such as closed system words (prepositions, articles, pronouns, conjunctions), which have a wide range of meanings. Additionally, the meaning of individual words is influenced by the context and the situation. The contextual values are not likely to be identical and the situations may vary to a greater or lesser extent.

Transposition

This is one of the most common procedures used in translation. It involves replacing a grammatical structure in the SL with one of a different type in the TL in order to achieve the same effect. I shall deal with this in greater detail below.

Modulation

This and transposition are the two main processes in translation. Modulation entails a change in lexical elements, a shift in the point of view. Transposition remains within the domain of the grammar book, but modulation is largely concerned with the dictionary meaning (strictly speaking, with an ideal dictionary). Transposition and modulation may take place at the same time; a transposition may also be a modulation. The procedure is dealt with more fully below.

Adaptation

This procedure is used when the others do not suffice. It involves modifying the concept, or using a situation analogous to the SL situation though not identical to it. An adaptation may at the same time entail modulation and transposition. It goes beyond language.

When there has been long-standing contact between two languages and cultures, certain translation equivalences become established. *Set theory* for *Mengenlehre* and similar conceptual borrowings are one type of equivalence.

Either utterance (SL or TL) in the translation situation may be optional or obligatory. From the translator's point of view, points that are obligatory in the TL are the most important.

When there are disparities between the SL and the TL message, as there nearly always are, the principle of compensation is applied. This means that if an emphasis in the SL cannot be reproduced in the corresponding position in the TL, it may be necessary to introduce an emphasis at another position in the TL. The translator must guard against the temptation to over-compensate, or to compensate unnecessarily, i.e. where the TL does not require the element that is obligatory in the SL.

die Geschichte der Müllerei reicht bis in *die Zeit zurück* ...
the history of the miller's trade reaches far *back to the time* ...

The SL has an element, *bis in* (**until into*), lacking in the TL. The compensation for this prepositional form in the SL is an extra adverbial element in the TL, which introduces an intensity into the sentence, though this is not quite the same as the intensity in the SL. A literal version would read:

The history of the miller's trade reaches back until into the time ...

This, however, would be over-translating, as 'until into' is redundant in this context. But 'to the time' would lack the emphasis of the SL.

LEVELS OF MATCHING

```
┌─────────────────────────────────────────────────────────────────┐
│ ┌─────────────────────────────────────────────────────┐         │
│ │ GRAMMATICAL                                          │         │
│ │ structure                        grammar             │         │
│ │ rules                            morphology          │         │
│ │ grammatical transposition        syntax              │         │
│ │                                                      │         │
│ │ - - - - - - - - - - - - - - - co-occurrence restrictions - ┘   │
│ │                                                              │ │
│ │ LEXICAL                                                      │ │
│ │ vocabulary                       collocation                │ │
│ │ dictionary                                                  │ │
│ │ lexical transposition (modulation)                          │ │
│ │                                                             │ │
│ │ - - - - - - - - - - - - - - denotation - - - - - - - - - - ┘ │
│ │                             connotation                       │
│ │ SITUATION                                                     │
│ │ concepts          history                                     │
│ │                   geography                                   │
│ │                   culture                                     │
│ │ adaptation                                                    │
│ │                                                               │
│ └───────────────────────────────────────────────────────────────┘
└─────────────────────────────────────────────────────────────────┘
```

TRANSPOSITION

Transposition can take place within a language as well as between two languages. Paraphrase is transposition within a language: 'I am going to town tomorrow' and 'I will go to town tomorrow'. There is no change in denotative meaning.

The two categories for transposition between two languages are obligatory and optional transpositions. Where paraphrase is possible, the form is optional (i.e. as between SL and TL). Obligatory transpositions are those that must be made if the structure of the utterance is to comply with the grammatical rules of the TL; optional transpositions are not imposed by the rules, there are alternatives for them. What is optional and what is obligatory will depend on context and, above all, on situation.

English word order generally demands obligatory transpositions. A German Comp+V+N (Subject) sentence would normally require an English N+V+Comp order. German subordinate clauses in which the verb occurs at the end (N+Comp+V) have to be replaced by English clauses with the complement at the end (N+V+Comp).

Transpositions belong to two principal classes: those that entail the replacement of one grammatical form by another; and those that entail replacement of a grammatical form by a lexical one, or vice versa. The translation of er trinkt ständig by he drinks all the time is a purely grammatical transposition – an adverbial phrase is used instead of an adverb. But the translation he is a chronic drinker is of the second type, because lexical means have been used to express the content of a grammatical item in the SL. If er erfindet eine neue Maschine is translated as he makes the discovery of a new machine, grammatical means have been used in the TL to express something that is expressed lexically in the SL. The lexicalizations and grammaticalizations are analogous in translation to paraphrase within a single language.

The purely grammatical transpositions comprise: change in word class (class shift); change in rank of item from word to phrase and vice versa (unit shift); change in word order (structure shift); and change of plural to singular (internal shift).[1]

Class shift

The four word classes that can be interchanged in translation are noun, verb, adjective and adverb. In all, twelve types of transposition are possible here: N – V, N – A, N – Adv; V – N, V – A, V – Adv; A – N, A – V, A – Adv; Adv – N, Adv – V, Adv – A.

The more common class transpositions are N – V (*bei der Analyse der Wärmeabgabe = when analysing the heat dissipation*); V – N (*die Untersuchungen . . . wurde . . . von Schlegel . . . summiert = a summary of the investigations . . . was made by Schlegel*); N – V (*dies gab uns die Möglichkeit = this enabled us*).

Many class shifts are optional and can be paraphrased in the TL; the translator thus achieves one-to-one grammatical correspondence with the SL: *in the analysis of the heat dissipation; the investigations were summarized by Schlegel; this gave us the possibility.*

But the following shift from noun to verb form is preferable to a more literal rendering: *zur Verbesserung der Bleiverflüchtigung = to improve volatility of the coal*. Similarly the shift from verb to noun in *es hat sich ergeben, dass . . . = the result was that* is obligatory, since a reflexive utterance is not grammatical in English (**it had itself resulted that . . .*)

Unit shift

This is a change from word to phrase, a word to a clause, a phrase to a clause and vice versa – a change in the rank of the unit. The German capacity for fusing lexical items to form one word often means that an English rendering has to express a single item by two or more items. The rendering of fused article and preposition in German entails the same type of shift. *Zur (zu + der)* expresses three items of information: direction or location in the preposition *zu*; the specification contained in the article *der*; and the case information in the article. At least two words are required to express this information in English, as well as word order, *to + the*. In practice, of course, a class shift obviates the need for this procedure, as with *zur Verbesserung der Bleiverflüchtigung*. Sometimes the TL requires addition. Thus *dazu* (*to this*) must normally be translated as *to this end* or *to achieve this*. But English tends to be more terse than German and reduces syntactic formations:

Die quantitative Berechnung der Bakterien wurde nach der mikroskopischen Methode *ausgeführt . . .*

Quantitative determinations of bacteria were carried out microscopically . . .

And again:

die auf Grund der Zahl der Bakterien berechnete Grösse
(**the amounts calculated on the basis of the number of bacteria*)
calculating from the bacterial number

This also involves class and structural shift (N – V and change in word order). Impersonal statements with *man* are usually rendered by *it + be* – a unit shift (*mann kann erreichen = it is possible to achieve*). *One can achieve*, a structural shift, is also permissible, however.

Structure shift

The necessary change in word order is possibly the most conspicuous feature of translation from German to English. It can be summed up in the formula Ab–Ba. The dominant term in an utterance becomes subordinate, or merely occupies a different position. In *Flugzeuge für den Kurzstreckenverkehr* (**aircraft for short-distance traffic*) = *short-haul airlines*, there has, among other things, been a reshuffling of grammatical elements (N + PpP = A + N). The qualifier following the noun precedes it in the TL.

> *Produktion der Biomasse heterotropher Bakterien und die Geschwindigkeit ihrer Vermehrung im Rybinsk-Stausee*
> (**production of the biomass of the heterotrophic bacteria and the rate of their increase in the Rybinsk Dam*)
> *production of the biomass of heterotrophic bacteria and their rate of increase in the Rybinsk Dam*

A similar reorganization – this time in the sequence of the possessive pronoun 'their' and the phrase it modifies – has taken place here.

German passive clauses have the general form Cn+S+Comp+P. The English parallel Cn+S+P+Comp:

> ... *so dass die Untersuchungen mit höheren als in der Praxis üblichen Konzentrationen durchgeführt werden konnten*
> ... *so that investigations may be carried out with concentrations higher than those normally employed in practice*

Internal shift

These are changes in voice, between transitivity and intransitivity, modality, degree of comparison, attributive and predicative use of adjective, plural and singular, definite and indefinite article, tense and so on.

Voice changes may be between passive and active or between either of these and reflexive forms. A change between passive and active is normally optional: *Untersuchungen wurden vom Romanenko (1964) angestellt* = (*investigations were launched by Romanenko (1964)*) or *Romanenko (1964) investigated*.

die Druckkräfte auf die senkrechten Flächen heben sich auf
the compressive forces on the vertical surfaces are cancelled out

Changes of this kind from reflexive to passive (or sometimes to active) are normally obligatory, because the reflexive form is rare in English.

Transitivity changes may mean a change from a simple verb form to a phrasal verb in English, corresponding to a morphological change in German such as the addition of the prefix *be-* to *lasten* (intransitive) to form the transitive *belasten* (to load). English verbs can easily change their function from transitive to intransitive without any change, and two optional forms are often available: *beeinflusst* can be translated either as *influences* or as *has an influence on*.

die Steroidchemie hat vor allem dadurch besondere Bedeutung erlangt, dass . . .
steroid chemistry has gained in importance, especially because . . .

This version is preferable to *has gained importance* or *has achieved importance*, but it is not obligatory.

The subjunctive form is fairly common in German, but rare in English. It is normally translated by the indicative, as in the following:

Abschliessend sei in diesem Abschnitt über das Wellenbild des Lichtes noch gezeigt, dass . . .
to conclude this chapter on the wave theory of light, it will be shown that . . .

The subjunctive forms of modal auxiliaries are rendered by a one-to-nought correspondence:

hierbei müssen gleichzeitig die physikalischen oder chemischen Vorbehandlungsmethoden geprüft werden . . .
at the same time the physical or chemical pre-treatment methods must be tested . . .

The form *können* may be rendered as could, but it tends to follow the above pattern: *diese Arbeiten können heute nur . . . durchgeführt werden = this kind of work can nowadays be carried out . . . only . . .*

The treatment of the article depends on the different ranges of use in the SL and the TL. Transfers include definite article to indefinite and vice versa; and article to omission of article and vice versa. This process can result in many mistakes, particularly with abstract nouns:

> *für die Reinigung, die Vorbereitung und die Vermahlung*
>   *with regard to cleaning, conditioning and grinding*

Change from definite to indefinite may be optional:

> *die zweite Erhöhung der Photosynthese wurde Mitte September beobachtet*
>   *a second increase of the photosynthetic activity was experienced during the middle of September*

The correspondences in the use of the article between German and English are, however, greater than the discrepancies.

Changes in number are also related to the use of abstract nouns. English uncountables rarely have plurals, but what may be regarded as an uncountable in English is not necessarily one in German. A number of nouns occur in the plural in German but in the singular in English (*Auskunfte = information, Erkenntnisse = knowledge*). But the reverse also applies (*Physik = physics, Matrizenmechanik = matrix mechanics, Mathematik = mathematics, 5 Yard = 5 yards*).

MODULATION

There are two classes of modulation: fixed or invariant modulation, when established translation equivalences are used; and free modulation, where an equivalent is created for the occasion. Fixed modulations are found in bilingual dictionaries and glossaries, but all modulations were originally free modulations. A typical fixed modulation is contained in the equation *es gibt: there is*. In the SL a lexical verb ('it gives') is used that has largely lost its lexical meaning for the speakers of the language, while in the TL a functional verb is put in the same position and performs the same role. The phrases are equivalent, not the verbs. In another context, *geben* will not be equivalent to *be*. All modulations are two-to-one,

one-to-two, one-to-nought, nought-to-one correspondences on the grammatical level. It is precisely because grammatical means are inadequate that modulation must be used.

A more complex example of a fixed modulation (though a very common one in technical language) is *die stufenlose Geschwindig-keitsregelung* = *infinitely variable speed regulation*. A great number of equivalents for technical terms are found to be modulations. This is particularly true of workshop and sales languages, which is only to be expected, since in each language we arrive at the term from the concept. The expression just cited is a fixed modulation because it is the accepted equivalent in the subject field, mechanical engineering. It is a modulation because the viewpoint has changed from a negative concept (step-less, i.e. not graduated but continuous) to a positive one. This process is possible because part of the range of meaning of the expression overlaps with part of the range of meaning of the English word.

Borderline cases of modulation are expressions such as *es gibt*: *there is*, which are virtually devoid of lexical meaning and can almost be considered to be grammatical forms. Other modulations approach the extreme where they merge with the procedure of adaptation.

The three ideas underlying modulation are:

inclusion, specialization (B-Bc)
           generalization (Bc-B)
opposition $(+ -)$ animate–inanimate
                     static–dynamic
                     positive–privative
                     cause–effect

The idea of lexical oppositions plays an important part in modulation

change of state   dimensions
                    directions
                    quantity–quality
                    symbols
                    sensations

The main modulations used in translation are listed below. A great many, it will be observed, are of the inclusion type (general/particular).

Abstract–concrete

This process is similar to the grammatical substitution of an abstract (collective) noun for a concrete (plural) form in German. A difference in level of abstraction can often be noticed:

Startimpuls	*switching* (c–a)
Öffnungsimpuls	*opening signal* (a–c)
Wiedereintrittskörper	*re-entry vehicle* (a–c)
Trinkmilch	*market milk* (c–a)

Part for whole

When a part represents the whole, the procedure is one of exclusion (Bc-B). The name of a capital city may represent a state (*Bonn's view on the matter*). A locality may represent an institution (*Wall Street*).

Aufgabenstellung	*problem* (instead of 'statement of problem')
in jahrelanger Praxis	*throughout many years of practical use* (here the relation is one of inclusion)
Luftschallschutz	*airborne sound insulation* (inclusion)
Prüfanstalt	*material testing laboratory* (inclusion)
mit relativ wenig apparativen Aufwand	*with little mechanical effort* (exclusion)

Animate–inanimate

This class of modulation has important collocational consequences, as we have seen; and there are often restrictions on the type of words that may accompany an animate or inanimate word.

Workshop language uses animate expressions for inanimate things on a large scale, and scientific language is by no means devoid of it. We are here concerned with the change of an animate in the SL to an inanimate in the TL, and vice versa:

das Sperren der Düse	*the choking of the nozzle* (inan–an)
die Zahl der zulässigen Fremdstoffe in Lebensmitteln steigt ins Unübersehbare	*the number of admissible impurities in goods becomes unlimited* (an–inan)

physiologisch gesehen	*from the physiological point of view*
	(an–inan)
Düsenblöcke	*throat pieces* (inan–an)

Static-dynamic

This is a common type of modulation:

gegen die Schallentstehung	*against the source of the sound*
die Festigkeit der Isolierungen	*the capacity of insulating materials*
gegenüber Schaltstossspan-	*to resist switching surges*
nungen	

This modulation is often accompanied by change of word class – noun to verb; verb to noun; noun representing an action to noun representing a state and vice versa; participle form to noun form and so on.

Positive–privative

This modulation is not merely negation in the grammatical sense but involves viewing a situation from the opposite direction to the way it is viewed in the SL. We can, for instance, say something is not there, but we can also say that it is absent. The first is negation, the second privation. *Loch, hole* are privative nouns; *nackt, naked* are privative adjectives (compare *unbekleidet, unclothed*, which are negatives); *mangeln, to lack* are privative verbs (compare *not to have*). If *daher fehlt es nicht an Vorschlägen* (**hence proposals have not lacked*) is translated as *that is why it has been suggested*; this is a modulation from privative to positive.

The typical morphological contrasts with *un-, -frei, -los* are negative–positive modulations:

Die unerreichten lötfreien Anschlusse …
… unique non-soldered terminals- …

Cause–effect

The event may be seen in the SL in terms of the cause but in the TL in terms of the result. Associated with this is the means-end modulation:

es stellt die beste Lösung des	*this is the best way to solve the prob-*
Problems dar	*lem*
denn die Hauptfehler sind	*the main defects invariably result*
immer konstruktiv bedingt	*from the installation*

To this type belong constructions such as *lassen erkennen* (*let know*): *indicate*; *die Antriebsleistung* (*the driving power*): *the input power*; *die installierte Leistung* (*the installed power*): *the available power*.

Dimension

Here a dimension in the SL is replaced by a different dimension in the TL (breadth by height, for instance *das damals erstmals in grösserem Rahmen verwirklicht wurde*: *which has been carried out for the first time on a bigger scale*). Changes in direction belong to this group. Prepositions play an important role in modulations of this kind. Prepositions have such a wide lexical range that they must be contextually determined.

The concept of 'strong' is frequently replaced by that of 'high' (*ein niedriger Wasserstand und starke Durchwärmung des Wassers*: *a low water level and a high degree of warming up of the water*). We cannot say 'strong warming', or even 'intense warming'.

A good example of dimensional modulation, which also involves a change from abstract to concrete, resulting in greater clarity, is: *das Abwasser . . . so an der Rohrwandung verteilt, dass es als Mantel abfliesst*: *the waste water is distributed over the wall of the pipe discharging as a column* (literally: *the waste water is distributed at the pipe wall so that it flows off as a cylindrical layer*).

Intensity

A change in intensity or emphasis, attenuation or intensification, is sometimes necessary. Thus *ausserordentlich verschieden* must be attenuated, since *extraordinarily different* sounds odd in English. A better translation, *entirely different*, is not quite as strong. The expression *aus diesem Grunde ergab sich als zweckmässig* is better translated *as for this reason it seemed useful*, while *Autotitrationsmethoden werden durch Esterasen gestört*: *autotitration methods are affected by esterases*.

Demarcations

These are changes in intervals and limits, as when *Begrenzungsfläche* is translated as *contact area* rather than *boundary area*. This also involves a reversal of point of view.

Metaphors, symbols

These form an extremely important class. We might cite as an example a chemical engineering term such as *Stromungsgeometrie* (*current/flux/flow geometry) = packing geometry or Schüttung (*discharge) = bed. There may be qualitative changes, as with bei einiger Überlegung (*with some judgement) = with a little forethought, or a change of metaphor: in sechs Stationen der Standardexpedition becomes in the six stations of the permanent expedition, and in legal language zur Veräusserung und Belastung von Grundstücken (*towards externalizing and loading pieces of land) becomes for alienating and encumbering properties.

ADAPTATION

This takes two forms. The first involves rendering the situation depicted in the SL, but with different lexical units. The translation of proverbs is an example: Altweibersommer (*old wives' summer) = Indian Summer. Varying historical experiences are reflected in these locutions, but other influences may also be at work. Further, a lexical adaptation of this type may also be a dictionary equivalent, as is the one quoted. It is the counterpart on another level of invariant modulation. Proverbs are in a sense the repository of national wisdom and they are very much bound up with a particular culture. Nevertheless, there can be overlapping, just as there is cultural overlapping. For example der Kopf in den Sand stecken corresponds to English bury (stick) one's head in the sand and both paraphrase in the same way (Vogel-Strauss-Politik treiben = pursue an ostrich policy). There is variation with zwei Fliegen mit einer Klappe schlagen (strike two flies with one blow), but it is easy to see the parallel in kill two birds with one stone. These expressions are more or less clichés and can often be found in dictionaries.

The second kind of adaptation is much more difficult. This entails the creation of a situation in the TL analogous to the one in the SL, because an identical situation does not exist in both languages. For example there is one-to-one correlation on all levels between Fussball and football, but German has no equivalent for cricket, since the game is scarcely ever played in the German-speaking world. To translate it into German would involve borrowing it completely as das Cricket. In some circumstances, however, it may be necessary to suggest a game with an analogous position in the German sports world. In that event it might be rendered by Handball, familiar to the German reader, rather than by, say, the circumlocution englisches Nationalballspiel ('English national ball game').

The correspondence may be nought-to-one or one-to-nought. The absent element may be a concrete object, a set of circumstances, an institution, a product, a relationship or a concept. Problems usually arise with the one-to-nought correspondence, but the TL may have a very compact and concise expression for something expressed rather vaguely in the SL, or with much circumlocution. For example the expression *englisches National-ballspiel* in the SL could be translated as *cricket* in the TL. The caption to a picture in a German technical journal *BO* 105 – *Ein Nahver-kehrsmittel* is rendered as *BO* 105 – *Executive Helicopter*. The German is more general ('short-distance means of transport') but the English is much more accurate on the sales language level. This is an adaptation. The English expression is independent of the German but it is related to the same extra-linguistic reality as the German.

This example illustrates the most difficult aspect of translation. But at the same time it illustrates the technique that should underlie all translation. Translation should be from concept to concept, from situation to situation, even if the translator's path is eased by correspondence on the grammatical and lexical levels. There are many snares on the path of these equivalences or seeming equivalences.

The expression *ein Normall-Diagramm* has a superficial resemblance to *a normal diagram, but reference to the actual object (which is tantamount to saying using one's knowledge of the subject field) will yield *a standard flow sheet*. Note that *standard* is an established translation equivalent of *normal*, while *Diagramm* need not necessarily be *flow sheet*.

Institutional differences exert, as we have seen, a subtle influence on the German means of expression as compared with the English, and if part of the meaning overlaps this should not conceal the fact that the correspondence is only partial, and that we are achieving only an approximation. This happens, as we have seen, when *Gewerbe* is translated as *trade*, because the concept of 'trade' does not have the same significance in German as in English (see p. 154). Similarly, *Wirtschaft* is generally translated as *economy*, although it has a much wider meaning. *Anlage* can generally be rendered as *plant* or *installation*, but it may also mean *contact*, *bearing*, *character* (endowment), *capital investment* or *design*, among other things. Is it going too far to suggest that the English equivalent for *Wettbewerbsbeschränkungen* (**competition restrictions*) is *restraint of trade*, because competition is taken for granted in the English-speaking world? The historical background would tend to confirm this, for the concept of freedom of trade is much more deeply rooted in English history than in German history. In the same way, German has *Jägerleitstationen* where English speaks of *fighter control stations* (aeronautics), because hunting continued for a longer time and on a wider scale in Germany than in England. Other influences also enter into the matter, as can be seen in Afrikaans *vegeskader* (*fighter squadron*). Big game hunting is still possible in South Africa, and Afrikaans follows the German tradition in *jageskader* = *pursuit squadron*; but Afrikaans is also strongly influenced by English, the other official language of the country, and therefore tends to construct its terminology on an English rather than a Germanic pattern. (It is a *Wurm* (*worm*) language, not a *Schnecke* (*snail*) one).

A word such as *Leistung*, which is in common use in engineering contexts, has a vast range. It may mean *efficiency, power, capacity, output, life, accomplishment, achievement, result, obligation* and so on. It must be defined either by a determinant element, as in *elektrische Leistung* = *electrical power*, or by the context or situation. When the term describes the performance of a machine in terms of energy transformation, its equivalent is *efficiency*; when it describes the work done within a time unit, it is *power*; when it is referring to oils, it is *serviceableness*; when referring to a human being, it means *efficiency, achievement* or *accomplishment*. Where the meaning has to be derived from the situation, we must use the procedure of adaptation of the second type.

Standarized technical terminology often exemplifies this procedure, since standardization implies proceeding from the concept to the term in each language. It is not a procedure on the linguistic level – from language to language – but from situation to language.

A common concept underlies the following pairs of aeronautical terms, for example:

Aufklärer	*Reconnaissance aircraft*
(*aufklären*	*clear up, explain, enlighten, inform, scout*)
Maschinenkanonen	*automatic anti-aircraft guns* (**machine cannons, guns*)

Radarsuchköpfen   *radar-tracking homing heads (*radar search heads)*

In the above examples the ideas are linked and there is partial overlapping. Reconnaissance involves exploring and examining territory in order to learn something about an enemy, and an *Aufklärer* provides enlightenment. Machines are largely automatic and cannons are large anti-aircraft guns. The German radar heads search; the English ones track (pursuit implies searching) and home (workshop language), which suggest finding a target or resting-place. The incomplete overlapping is not so important here because a term is in part only a label and therefore need suggest the concept only in part. It has its place in the terminological system and the subject field therefore completes its meaning.

CHAPTER 14

Adequacy

🔲🔲🔲🔲🔲🔲

JUDGING the quality of a translation and establishing a standard for a satisfactory piece of work is always a vexing problem. But it is a very important matter in technical translation, where the question of economics must be very much in the forefront of the translator's mind. What both client and translator want is a satisfactory translation achieved with the minimum expenditure of time and effort. But what does satisfactory mean? No one yardstick is available for all translations under all circumstances. The adequacy of a translation varies according to circumstances – the needs of the user, the subject matter, the type of language used and so on. The needs of the user must obviously be the decisive consideration.

As with any other product, it is uneconomic to spend more time and energy than is strictly necessary on the process of production; but it is self-defeating to spend too little time and energy, and so produce something inferior. Inferior in this case means a work that is unreadable, or difficult to read, and possibly also inaccurate; in these circumstances the burden has been shifted from the shoulders of the translator to those of the reader. Since a technical translation often has more than one reader, this shift of onus may be repeated several times.

Readers' needs; the level of the text (its difficulty); the nature of the subject; the type of document – all these factors vary. One client will simply require a general indication of what the text is about, before he decides whether to have it translated or not; a second will want only a paragraph. The material in question may be a newspaper article, an abstract, a letter, a paper describing an experiment, an article in a learned journal or a book. Some items have only an ephemeral interest; others will be read again and again over many years. The standard required will vary accordingly.

In any event an adequate translation will always be one that has been produced with just enough expenditure of time and energy to meet the needs of the consumer. It should not be of a higher quality than he requires if this will involve a higher cost. But it must not be of a lower quality either, if this means that the reader will have to spend much time and energy in deciphering it. The technical translator sets his sights on adequacy, not on perfection.

Three factors determine adequacy: accuracy (the translation must convey the information contained in the original with as little distortion as possible); intelligibility or readability (the reader should not have to struggle to work out what it is all about because it has been expressed badly by the translator); speed (the client requires the finished work as soon as possible).

Each of these requirements conflicts with the others, and each varies according to circumstances. An ideal translation is perfectly faithful to the SL text, is perfectly intelligible and is produced within a very short time. The closer the TL version comes to the SL text, the greater the fidelity, but it is not necessarily more intelligible. Fidelity on the grammatical level may result in obstacles to understanding, whereas a full situational translation may be required, demanding much time and energy.

The relationship between accuracy and intelligibility is not arbitrary; ideally an accurate rendering should also be an intelligible one. But the original may be poorly expressed, or it may be expressed in a manner appropriate to the SL but not to the TL. The demand for speed conflicts even more strongly with the need for accuracy and intelligibility. The greater the speed, the more the standard of accuracy and intelligibility will suffer. At the best of times a text will tend to become distorted when it is transferred into another language. The translator may aim at readability above all and be less concerned with accuracy. Alternatively, he may be more concerned with rendering the original faithfully than with the readability of the TL text. There is always a tendency in translation for the SL to interfere with the TL expression, and for the translation to retain characteristics of the original language. Speed will aggravate these hazards.

We may sum up the problems as follows: a communication process will inevitably undergo 'entropy' (the longer the chain of communication the greater the deterioration in the message). The more links there are, the more stages there are between sender and receiver, the greater the probability of loss and distortion of information. The process of translation obviously introduces additional stages into the transmission of the message. The most accurate technical translation is one that diverges from the original text only to the extent necessary for intelligibility. Keeping close to the original may impair readability. Speed will increase the possibility of distortion and lack of intelligibility.

An adequate translation is always a compromise between conflicting demands. It tends to concede fidelity to the requirements of intelligibility and of speed, but the extent of the surrender will depend on the circumstances.

Translations can be graded in the following terms:[1]

rough translation, e.g. word-for-word
This is low on intelligibility and imposes a high communication load on the reader, but scores high on speed.

grammatical translation
The necessary transpositions are made so that the translation conforms to the grammatical rules of the TL. Improved intelligibility, less speed.

lexical translation
Not only have the necessary transpositions been made but the words have been chosen to conform with the usage of the TL. Intelligibility even higher, speed still lower. This grade of translation will also tend to be more accurate – this is partly what a correct choice of words means.

stylistic translation
The transpositions and the lexical choices are made but register (scientific, workshop, consumer etc.) and level are taken into account. Intelligibility high. Accuracy can be high. Even less speed.

The above grading is therefore in terms of increasing intelligibility and decreasing speed. Increasing accuracy is also assumed, but this may not be a substantial factor – and there may not be any improvement at all. A scale of increasing accuracy could similarly be made against decreasing speed. A minimum level of accuracy and intelligibility must be presupposed. In the majority of cases, a version that is at least grammatically correct must be presupposed.

Accuracy levels can be graded along the following lines:

1 general idea of original conveyed

Enough to indicate, for example, whether information is relevant or irrelevant and whether the translation worthwhile. This is not always a simple matter, and unfortunately sometimes it is impossible to know this until a full translation has been made.

2 information content conveyed sufficiently for action to be taken (e.g. a letter)
3 most of the information conveyed (say, 75 per cent, though quantities are very deceptive)
4 virtually all the information conveyed

Generally only the fourth grade of accuracy is satisfactory.

Many attempts have been made, and will continue to be made, to devise objective methods for establishing the quality of a translation. Attempts have been made to assess the degree of distortion in a translation by means of a process known as 'back translation'. In its simplest form this involves having the TL text translated back into the SL and comparing the result with the original, let us say SL" compared with SL'. The two extremes on the scale we can derive from this method occur when SL" and SL' correspond perfectly and when it is impossible to see any relation at all between SL" and SL'. Between these two extremes, degrees of adequacy are established.

One experiment involved a translation from English (1) to French (2); another from French (2) to English (3); another from English (3) to French (4); then finally from French (4) to English (5). In all, four translations were produced, and at each stage a different translator did the work. The results were not conclusive, however.[2] Of course the inevitable element of subjectivity in the comparisons is difficult to eliminate.

A more systematic version of this method[3] established the following scale:

1 perfect identity of meaning
2 close similarity in meaning
3 recognizably similar but with substantial change of meaning
4 more different than similar in meaning
5 no similarity at all in meaning

ADEQUACY

The criteria used here are still subjective, and depend on the opinion of individual translators. Moreover, demands will vary according to different circumstances, as we have already seen. 'Identity of meaning' is very difficult to establish, not least because of the many links in the communication process that I mentioned earlier. Consider a situation where a recorder is attached to a telephone to inform callers of a change in number. The listener hears the following message: 'The telephone number of Mr X is now 1234.' This is perfectly clear and the message will be conveyed as intended. But if instead the recorder says: 'The telephone number of Mr X today is 1234', a problem arises. When is 'today'? If 'today' means the same as 'from now on', there is no difficulty; but if the reference is to one particular day, how is the listener to know what the given day is? We know too that pronouns, place references, word order and so on may be ambiguous. Ambiguities of this kind are by no means unusual. They occur constantly in everyday life. And identity of meaning always is assumed.

This corresponds to the principle in information theory which says that the more restricted the possibilities of choice in the message, the lower the information content. This in turn means a proportionately greater likelihood of coincidence between the sender's intention and the reader's interpretation. If the possibilities of alternatives for 'today' are to be restricted, additional information is needed. This may take the form of the listener's advance knowledge (he was told the day before that the number would change to 1234 on the following day); the listener's knowledge that the sender has the habit of using 1234 on a particular day of the week, and so on. The information may be given in the following form: 'The telephone number of Mr X today – 1 January – is 1234.' If the listener knows that this is the first day on which the message has been given, he can be reasonably certain about what 'today' means. Further, the importance of knowing exactly what 'today' means will vary. It may be a life-and-death matter, or tomorrow or the next day may do just as well. The criterion for 'identity of meaning' is likely to change accordingly. But the rougher the criteria, the more possibility there will be of agreement between a number of people. Here again a compromise is necessary – this time between the conflicting demands of accurate criteria and the demands of agreement between a number of people on the criteria.

In practice a grading that corresponds to the different levels of translating outlined – word-for-word, grammatical, lexical, situational – will probably be made. It is generally relatively easy to distinguish between say, three grades of quality: rough translation, working translation and publishable translation.

Rough translation

This provides information on whether it is profitable to make a better translation or not. It must give some idea of the denotative content of the message, but it may offend against grammatical rules and the rules for collocation of lexical units.

Working translation

This may give the reader who is a specialist in the subject enough information for his purpose. It may also be raw material for revision. It should communicate denotative content and must obey the obligatory grammatical rules of the TL. All the necessary transpositions will have been made. It may, in addition, give the designative meaning (value) of the words, performing the required modulation.

Conceptual or situational translation

This communicates denotative content; obeys the obligatory grammatical rules of the TL and the lexical collocation rules; communicates the values (designative content) of the lexical items and the pragmatic meaning. This last term means that the reader is given the information needed for performing operations that a native speaker of the SL would be able to do from the original text.

A translation available to the general public and listed in an international index of published translations will read like an original work and not like a translation at all.

This grade of quality corresponds to translation on the level of situation, in which the necessary transpositions, modulations and adaptations have been made. The idioms and proverbs of the TL are used where required, and compensatory changes are made for untranslatable situations.

A rough translation may be made to enable the potential user to decide on whether he wants it at all.

A working translation may be made by a staff translator for a research worker in the same organization. It may provide confirmation, which a rough translation may not be able to do, that he is not duplicating work done elsewhere. It may be requested as better than nothing when speed is essential. It may be used on a temporary basis, pending the arrival of a publishable translation. A rough translation, it should be noted, does not always provide sufficient information on which to base a decision. Even a working translation may turn out to be inadequate here. It depends on what information is required – a date; the address where a conference will be held; a detail in an experiment; the context of an equation; the explanation of a hypothesis; a description of the operation of an engineering plant. Quality in technical translation is a function of who is saying something, what is being said, who wants to know, for what purpose does he want to know.

The process of translation can also be viewed in terms of these grades of quality. The first result of a translator's efforts will often be a rough version and may not be written down. It may exist simply in his mind; or he may merely list important points. At this stage he is establishing the vocabulary and terminological equivalents. In a fairly large translation unit this task may be undertaken by a terminological specialist who does nothing else.

In practice, a translator does not work in as compartmentalized a manner as the above account may suggest. He will be performing other operations on the text at the same time – grammatical and lexical operations. But it is undoubtedly more efficient to establish terminology first of all. In addition the translation may be given *viva voce*, the translator rendering it at sight to the user. In that event, he may pay little attention to grammar.

If a working translation is required the rough version will be polished accordingly. But the working translation may be the first result of the translator's efforts. It will perhaps be handed over to the user for immediate use, for temporary use until a better version is available or for correction of terminology and general comment. These purposes may of course be combined. The user may find that this standard of translation is all he requires. Alternatively the work may be given to another member of the translation staff who specializes only in editing and revising translations, sometimes working on a version corrected by the user. The editor, who should preferably not be the original translator, prepares the final publishable version.

The stages have been divided up as though each was carried out by a different person. This may be the case, and it is desirable that it should be. In a smaller translation team, however, all these operations may well be carried out by a single person.

This is very much a rule-of-thumb method for assessing quality of translations, but unfortunately it is the best we have at the moment. An analysis of the ideas of intelligibility, accuracy and speed in translation will throw more light on the question of translation adequacy.

READABILITY

The intelligibility or readability of a text depends on the extent to which a) the reader's attention is aroused; b) his attention is maintained; and c) he remembers the content.

His attention is aroused if his expectations in the situation are fulfilled. A text by a research chemist on an experiment he has conducted will arouse the interest of another chemist who is working on the same theme. A text by a botanist will normally not have the same effect. The relevant elements are: the situation (scientific research, technological development or application, sales literature); the subject field (chemistry, physics, geology, mining); a particular topic in the subject field. In practical life it is a matter of stimulating a preference for one thing amid a host of competing interests. The potential reader is often very busy and harassed; he is probably very interested in his work, and at any rate he wants to make headway, both financially and in status and prestige. He is swamped by literature, and he is working on particular problems. His attention will be held only if the text is relevant to his own interests and needs. A striking title for a work of science popularization may capture the attention of the general public, but in the general run of technical papers the title is solely an indication of contents. It may even carry a Universal Decimal Classification (UDC)† number to signify its topic accurately.

Brevity is important in a title, though national preferences vary. English titles, for instance, are generally shorter than Italian ones, although there is a growing tendency for conciseness in all Western languages.

The important thing about the title is that it will create expecta-

† The Universal Decimal Classification system is a numerical classification of books and documents in terms of subject matter. It is an international system and is used chiefly in scientific libraries.

tions. These expectations are dependent not only on the title, but on where the title appears. In a journal devoted to aeronautics a title containing the word *feedback* will create different expectations from those it would arouse if it appeared in an electronics journal. Different expectations will be created according to whether the journal is one concerned with research, with development, or with manufacture. The reader will expect to find a certain type of content; he will expect it to be presented in a certain way; he will expect a certain language to be used and so on.

The title will arouse his attention if it appears in a certain journal. It will tell him that the text deals with matters that are relevant to his own work.

The maintenance of interest depends largely on how the expectation is fulfilled. In the first place this depends on the information contained in the article. It may be relevant and significant throughout; in parts, or not at all. If it is not relevant and significant, the reader will pay no further attention to it, irrespective of anything else. If it is relevant and significant, this alone will hold his attention, but he may be irritated and frustrated because of poor presentation, or because the level of generalization is not appropriate. In a journal devoted to application he is not likely to be interested in a large amount of theory. A very abstract text will not usually be suitable for technologists or men in the workshop. The amount of information, too, must not be excessive for the type of text (workshop, research or development). A very abstract text organizes a large amount of information within a very small space. A very concrete one contains much less information within more space. A text on too high or too low a level of abstraction for the situation will tend to lose the reader's attention.

The next factor claiming the reader's attention will be the logic and consistency of the argument. It will not be held if the argument is seen to be absurd or badly presented.

The content will make a more than passing impression on the mind of the reader, if his expectations have been fulfilled and the article is found to be both important and relevant. It conveys information of value; it is at the right level of abstraction; and the argument is logical and consistent. A consistent theory that covers all the necessary information will be more memorable than data that do not fit into a logical and consistent scheme but appear as isolated fragments. The framework may be provided by the discipline or branch of the discipline, or it may be provided by the author of the article.

Formal features affect the reader's attention more than they contribute to the permanent impression made by the statement. The empirical rules for good technical style are largely structural. They are – according to some evidence – mnemonic devices that enable us to recall an earlier part of the utterance at a later stage.[4] Or they may act as signals to sustain flagging attention. Words like *wirklich* and *zwar* in German or *indeed* and *undoubtedly* in English, have this sort of function.

Inappropriate lexical items; collocations that do not conform to TL usage; items that overlap between the SL and TL but are used in the SL in a part of the range that does not occur in the TL – all these create barriers to attention. They find expression in repetition and verbosity. Use of the wrong register will result in jargon, as sometimes occurs in advertising.

Sentence length is one of the structural factors that affect intelligibility. Shorter sentences are easier to remember than longer ones and in technical language the tendency is to prefer short sentences. But a succession of short sentences is monotonous and fatiguing. The number of clauses and the manner in which they are linked will affect the reader's ability to remember them. German is more addicted to complicated sentence structure than English and in many cases an English reader would not tolerate the load carried in German. This is where something more than a working translation becomes advisable.

Simple sentences can also be difficult to carry over into another language if the thought they contain is novel and the terminology new. This is a very common aspect of technical translation.

Nested and self-embedded structures are obviously more difficult to remember than right-branching and left-branching ones. Nested structures involve clauses enclosed within a sentence:

1a *In einigen Fällen wird das zur Verfügung stehende Frischwasser, welches auch als Zusatzwasser für den thermischen Kreislauf benötigt wird, ausreichen, um . . .*

1b *In certain cases, the available fresh water which is also required as make-up water for the thermal cycle suffices for . . .*

Clauses of this kind do not occur very frequently in technical prose and self-embedded sentences on the pattern $S' + wh- + S''$

+ *sh*- S′ ″P′ + P″ (*the man who the people who wrote the letter spoke to . . .*) are even rarer. Left-branching structures are more popular: *John's brother's father's uncle.* In German the form is more likely to be a long compound: *die Turbinenauslegungsleistung* (**turbine's design's capacity*) = *planned turbine capacity.* Technical English will have a string of nouns in succession without any indication of possession; or English will turn the chain of participial, noun and adjectival modifiers and noun into a finite clause, as in *die benutzten Boltzmannschen Entropie-Wahrscheinlichkeits-Beziehung* (**the utilized Boltzmann-type entropy-probability relation*) which is transposed to give *the Boltzmann entropy-probability relation was employed.* This makes for greater clarity but is less condensed. The right-branching structure takes the form: *This is the cat that caught the rat that stole the cheese.* Although there is a marked tendency in technical writing to convey as much information as possible within the most compact space, this is generally expressed by means of terminology, with one term standing for many words, rather than by means of very involved clauses. A highly condensed style is quite often found in sales language and journalism, apparently so as to create the impression of dispensing with inessentials.

In general complicated clause structures are not found in technical language, though they do occur in texts using a good deal of legal and administrative phraseology.

Redundancy is an important factor in the matter of capturing and holding the reader's attention. An over-economical (over-concise) style is tiring to read and hard to follow. This is true of texts with a very high level of abstraction, which contain a large amount of information within a small compass, or with long and complicated syntactic constructions. Redundancy alleviates strain on the reader's attention and helps him to remember the text.

The TL text will usually contain more redundancy than the SL because more explanatory material must be incorporated. The redundancy required varies for different purposes and registers. It also varies from one language to another. It is a sort of guarantee against misinterpretation and inaccuracy. A text will generally contain more information than is needed. It is well known that a message that omits, say, the vowels, will still be intelligible. It is possible to recognize that *Bwr f th dg* on a garden gate means *Beware of the dog.* To write it in full is to introduce redundancy. German sentences have a high degree of redundancy – they include gender, case and number markers in both subject and object, when it may well be enough to have these for only one item. The existence of redundancy is recognized by ordinary speakers when they make free use of abbreviations like *i.e.* or truncated forms like *veg.* Excess, deficiency or appropriateness of redundancy are also criteria for assessing the efficiency of an utterance.

One language will require a more formal style than another. Italian is prone to more elevated styles of utterance than English (*tali tipi litologici = these rock types*). Similarly German often adopts a more formal tone than English and literal transfer sounds over-inflated in the TL. Then, too, the level of abstraction in German may be higher than that required in English: *Messwerte* (**measurement values*) = *readings*. German tends to be more abstract than English:

2a *Apfelsaft, physikalisch haltbar gemacht, verzehrfertig* (**apple juice, made physically durable, ready for consumption*)
2b *ready-to-serve preserved apple juice*
3a *Umfang und Stand der Arbeiten am Weltlebensmittelbuch* (**extent and state of the work on the world food book*)
3b *progress on Codex Alimentarius*

Failure to adjust the style makes for difficulty in reading.

ACCURACY

An 'accurate' rendering of 2a would be intelligible but clumsy and would take more time to read than 2b. An accurate version of 3a would also be intelligible, but clumsier than 3b. Should the translator risk accuracy for the sake of a statement that reads smoothly? In some circumstances readability will have to suffer, but the whole question of accuracy calls for closer inspection. An accurate rendering can be considered to transfer the content of the words (which includes designation as well as denotation), in which case the starred versions in parentheses are more accurate. But we are concerned with information, and from the point of view of information the expressions are merely labels for concepts or things. The expression 2a describes the type of apple juice and the correct description of this apple juice in English, in accordance with English usage, is 2b. If there were no English expression it would probably be necessary to sacrifice readability in order to convey the idea better. But since an English label does exist it is appropriate and accurate to use this. The same can be said of 3a. The concept is accurately given in English terms in 3b.

However, the continual creation of new terms for new concepts, products and processes is a fact of technical language. Often there is no established equivalent in the TL, and the item does not even exist in the TL culture.

4a *Grundsätze für Gütezeichen*
4b *quality mark principles*

This example is extracted from a paper published in 1956 in *Gewerblicher Rechtsschutz und Urheberrecht*, the organ of the *Deutsche Vereinigung für gewerblichen Rechtsschutz und Urheberrecht*. The name of the organization presents difficulties and in this case the translator had opted for accuracy rather than elegance: *German Association for Legal Protection in Trade and Industry and Originator's Rights*. The idea of 'quality marks' is also difficult to convey in English, and in a very real sense the entire article is an explanation of *Gütezeichen*. The version *quality mark* was regarded as being nearest conceptually to the original, different enough from *trade mark* and having the notion of higher quality than, say, approval by a Bureau of Standards. It is a very literal rendering and perhaps not the best term for the purpose.

The translator sometimes introduces new terms into the TL in this way. But it may happen that later a better term is coined, when the concept or object has become more familiar. At that stage the demand of intelligibility has made itself felt. An expression like *sputnik* or *Blitzkrieg* rolls off an English tongue easily enough and associates itself with a concept in the English-speaking word, but it is not as intelligible as, say, *feedback*, which is transparent to the English speaker.

SPEED

Not much need be said here about speed, as this requirement is self-explanatory. The user of a translation will invariably want it as soon as possible. To work under conditions of haste is not conducive to accurate or readable translation. The speed with which a translation can be made will depend on its nature, its subject matter, the language in which it is written, the level of abstraction and so on. Because of the fluctuating nature of these conditions, because of the many variables in the process, it is difficult to make any quantitative statements about speed of translation.

DIFFICULTY

Allied to these questions is the degree of difficulty of a translation. This depends on specific linguistic and non-linguistic considerations. The linguistic difficulty is related to the distance between the SL and the TL. German is an inflexional language with a developed system of case endings, verb conjugations and rules of sentence concord; it has an individual vocabulary, diverging from what may be called Western European vocabulary. English is classed as an analytical language with few inflections, and it shares to a large extent the common Western European vocabulary. In many ways it is closer to French than to German. But English is very much closer to German than it is to Tibetan or Chinese. A scale of Germanic languages would certainly have English at one end and German at the other, with Dutch in between. In general there is a greater translation distance between English and German than between English and French – partly as a result of cultural interchange over the centuries.

Cultural interchange is a very important factor. Even languages that are linguistically far apart may become closer from a translator's point of view if they have exercised a mutual cultural influence over a long period of time. These influences are especially pronounced in countries where bilingualism is the rule, although one language will tend to dominate the other in such cases. These contacts result principally in lexical borrowings, and to a lesser extent in grammatical effects. If there has been a great deal of translating activity between the two languages, many translation equivalents will have become established. Translation routes have been developed and form a model for later translations. It is easier to translate under these conditions than between two languages that are confronting each other for the first time.

The degree of difficulty is also affected by the subject matter. The nature of the discipline is important here. Certain sciences and technologies are advancing faster than others and consequently yield a bigger crop of neologisms. An entirely new field falls into the same category. If it is science-based, as with nuclear engineering, it is likely to have a better organized terminology.

Scientific material will be more difficult conceptually, being more abstract, but it will have more standardized terms and will tend to be better written than texts on other levels. Workshop material will not only be more concrete, have more colloquial elements and be conceptually easier to deal with, but there will be products and processes in the external world to which one can refer in order to interpret the utterance. Sales language, because of its emotive content and the semantic vacuity of many of its expressions, can be very awkward to translate.

Degree of difficulty is also dependent on whether the text is an article in a periodical, a specialized journal, a theoretical journal, a textbook, a worksheet, an instruction leaflet, a letter, a patent specification, a standard or an abstract. A high degree of accuracy is necessary for a standard, but it must also be convenient (intelligible). A patent also demands a high degree of accuracy, but of a different type – it must conform to very stringent legal requirements. The intention may well be to disclose as little information as possible consistent with those requirements.

The greatest difficulties in translation are caused by lack of an appropriate concept (institution, object) in the TL. The next group of difficulties is due to lexical disparities (e.g. value). Grammatical disparities are the least important here.

A scale devised by members of the German army's linguistic research group[5] shows an ascending order of difficulty:

relatively simple texts; popular science newspaper articles; simple technical instructions

average difficulty, translatable with conventional available resources and after some study of the material

difficult technical texts on new developments and theories on which there is little TL literature or virtually no terminology in current technical dictionaries

To these may be added lack of context, as when a book title has to be translated – this often happens if a translation service is also associated with a library – or a letter.

ERROR

The major errors that can occur in the process of translation are:

loss of information
lack of intelligibility

interference between sl and tl
incorrect level
errors in use of tl

Loss of information

The text may be inaccurate or false. All the information in a text is not of equal value, and a translation below publication standards may meet the needs of a particular reader. Inaccuracy may result from too free a translation and too much individuality. Technical style is generally impersonal and standardized.

Organizations have house styles and these tend to vary little from one organization to another. Most of the books on good technical writing give the same sort of advice and, as a result, styles will be similar. This impersonal style is more easily developed than an individual style. It meets the needs of technical communication. Introduction of individual expression may distort information.

Lack of understanding or carelessness can cause information errors – false information conveyed, an important paragraph or sentence left out.

Lack of intelligibility

The content may be transferred but in such a way that it requires effort on the part of the reader to understand it properly. Individual intrusions may be responsible for lack of intelligibility. Idiosyncratic use of language, even if it is not incorrect, may upset the expectancies of the situation.

Interference

This is the occupational disease of the translator. Usages peculiar to the sl are transferred into the tl. The mark of a timid and inexperienced translator is too close an adherence to the sl in style, vocabulary and word order. A notorious example involves succumbing to the blandishments of 'false friends' (*faux amis*), expressions that look alike in both languages but have different meanings. Fatigue and consequent carelessness may also cause this type of error. The translator must steer a middle course between the Scylla of literalism and the Charybdis of departing too far from the original.

Incorrect level

This involves replacing an SL utterance with the wrong level of abstraction or the wrong style in the TL. This may result from lack of sensitivity to the range of registers in the TL, but it can also arise from inability to discriminate between the styles in the SL. Knowledge of the TL levels is most important here.

Errors in use of TL

These include incorrect spelling, incorrect capitalization, inadequate punctuation, lexical errors, omissions and inaccuracies.

Faux amis are the best-known type of error in translation literature. The adjective *physikalisch* looks like *physical* and translates as *physical* in *physical science*, but it is ambiguous in other context such as *physical culture* (*culture of the body*), *physical effort* (*bodily effort*), *physical education*, *physical geography* and so on. False friends with affixes are especially tricky. *Physikalish* may give *physical*, *politisch political*, *mathematisch mathematical*, but *ökonomisch* is not necessarily *economical* – it may be *economic*. The phrase *ökonomisches Prinzip* is *economic principle*, the principle whereby the maximum can be achieved for minimum expenditure (*Wirtschaftlichkeit*). It is the adjective from *economy* as used in *political economy*, or from *economics* (*economic science*), whereas *economical* is synonymous with *thrifty* (*sparsam*).

In many contexts *dankenswert* will be equivalent to *gratifying*, but this would be a wrong choice in the following: *in dankenswerter Weise . . . hat die FAO . . . versucht . . .* (**in a gratifying way . . . the FAO undertook . . .*) *the FAO laudably undertook . . .* This is an example where carelessness, or more likely, inadequate command of the TL, is the source of the error. In this context *gratifying* is rather fulsome, and even *laudably* is open to question. A better version might run: *The FAO deserves credit for for having undertaken . . .*

Confusions may take place even on the phonemic level because of the similarities between many German and English forms (*Vater, father*; *Wasser, water*). Confusion may be due to lack of knowledge, probably of the SL in this case. Fatigue, possibly the worst enemy of the translator, may also be a culprit.

Chemical nomenclature deserves special mention as an area full of pitfalls. Here again the similarities encourage one to neglect the differences, to fall unsuspectingly into the pit. Some of the rules are delightfully simple – usually when German uses an international form:

Methan	*methane*
Propylen	*propylene*
Octan	*octane*
Äthylen	*ethylene*
Vinylchlorid	*vinyl chloride*

As soon as German resorts to national names, as we have seen earlier, the problem becomes more difficult, though we do know that the termination *-säure* is translated as acid:

Zitronsäure	*citric acid*
Bernsteinsäure	*succinic acid*

Another species of interference, and a source of false friends, occurs between different subject fields. We have noted the existence of polysemy across subject fields (plasma): this can lead to the use of a term in one field with the same meaning as it has in another. *Achse* can mean *axis, axis of symmetry* (*mathematics*), *axis of a lens* (optics), *axes* (lines of reference, crystallography), *axle* (technology), *axis* (botany), *axes* (zoology). This is a relatively harmless example since the range of the English term is more or less the same as that of the German. But *Durchlasswiderstand* means *forward resistance* in electronics, whereas in the construction industry it has other meanings:

Wasserdampf-Durchlasswiderstand	*water vapour transfer resistance*
Wärmedurchlasswiderstand	*overall thermal resistance*

The use of the electronics term will here elicit a blank, if not wild, stare from the reader.

Conditional clauses may also cause interference: *betrachtet man diese vielfältige und komplexe Einflüsse . . . dann wird klar, dass . . .* could be translated **if allowance is made for the interaction of these various and complex effects, it is shown that . . .* This is clumsy, though not unintelligible. The following version would be better: *these various and complex effects show that . . .* The first translation is too literal, as it has carried over the conditional form, which does not convey essential information. The *if . . . then* structure is merely stylistic and the straightforward declarative sentence is much clearer in English.

A common error is to translate *bei* as *in the case of*, when plain *in* will usually do: *bei feuchten Stoffen = in damp materials.*

CHAPTER 15

Dictionaries

੩੩੩੩੩੩

THE dictionary is the reference book that is most generally associated with translating. It might even stand as a symbol for it, but it is an instrument to be used with caution and discernment. Many people regard dictionaries as infallible authorities on language usage, but this is not the view of lexicographers. For the lexicographer, the dictionary records and describes usage; it does not establish it. Yet this is not entirely true. There are labels in the standard monolingual dictionaries like 'slang', 'colloquial' and so on, and even in *Webster's Seventh New Collegiate* we find the expression 'substandard'. These labels are not only descriptive: they also carry social connotations. Certainly they will be viewed by the general public as deprecatory, regardless of the dictionary maker's intention. Translation tends to make one more sophisticated or more sceptical about such things. From the translator's point of view, it is wise to remember that a dictionary is always out of date; that many of the expressions it records are no longer in common usage, that expressions referred to as colloquial or substandard may have risen into more formal use; that new expressions have come into use but are not yet recorded. This last group is the most common.

The dictionary records many things we know or think we know. It also records usages that a normal speaker of the language is not likely to know, to encounter or to use. But it omits a number of usages on which he requires guidance. It has limitations but, if used intelligently, it is of tremendous value, indeed indispensable.

We often talk simply of 'the dictionary', but in fact many different kinds of dictionaries exist, each with different uses. The first distinction to make is that between dictionaries and encyclopedias. The encyclopedias will be discussed in chapter 17. One type of dictionary shades off into the encyclopedia; and encyclopedias, like dictionaries, are arranged in alphabetical order. But a dictionary is primarily concerned with information about the behaviour of language, whereas an encyclopedia is concerned with knowledge of the world in general, which includes language.

Dictionaries in only one language are known as monolingual dictionaries. Bilingual dictionaries provide a list of words in one language with their equivalents in another language. Multilingual dictionaries provide equivalents in a number of languages.

All the entries in a monolingual dictionary are in one language. The great *Oxford English Dictionary* (OED) and the *Shorter Oxford*, *Webster's Third International*, *Webster's Seventh New Collegiate*, *Littré's Dictionnaire de la langue française* and G. Wahrig's *Das grosse deutsche Wörterbuch* are monolingual dictionaries on a similar model.

Other dictionaries include specialized vocabularies of a particcular social group, a geographical area or a vocation; dictionaries of slang or dialect; technical dictionaries; etymological dictionaries, which trace the origins of words; synonym dictionaries, which give the range of synonyms associated with a word; and analogical dictionaries, which are based on conceptual associations. The analogical dictionary is typified by the famous *Roget's Thesaurus* and other works on the same model such as Dornseiff's *Der deutsche Wortschatz nach Sachgruppen*.

The range of bilingual dictionaries is similar to that of the monolingual group and may include general or specialized works.

Multilingual dictionaries are usually specialized.

The monolingual dictionary ranges from the encyclopedic dictionary to a list of terms and equivalents at the end of a book, or printed separately. *Le Grand Larousse Encyclopédique* is an encyclopedia or a dictionary, depending upon which way you look at it – it is advertised as containing the entire language of France, with 450,000 entries. Encyclopedias are primarily concerned with the denotative meanings of lexical units such as words, but almost all dictionaries have encyclopedic elements. Dictionaries like Webster with illustrations are good examples of this. The historical dictionary of which the *Oxford* is a model is regarded as a repository of virtually the whole vocabulary of a language. This contention may be credible for the *OED*, but it is obviously not so for a dictionary like the *Wahrig*, which has only 100,000 entries.

The smaller monolingual dictionaries are for the most part condensed versions of the standardizing type of dictionary, i.e. the dictionary that is regarded as setting the standard for usage, though it does not necessarily set out to do so. These dictionaries are compact; omit many types of entry (obsolete and occasional applications) and are generally more up to date than the parent dictionary. They are published either in a medium size, say over 40,000 words – the *Petit Larousse* has some 70,000 words – or in a small size with about 10,000 words.

Ideally a dictionary should be an instrument of semantic discrimination. It should enable the user to choose between words for a given application; and it should offer information about the position of a word within a series and its value within a lexical structure. But this function is performed crudely by existing dictionaries. The crudity results from the arbitrary arrangement of items from the linguistic and conceptual points of view. The dictionary presents the language as an inventory, a list of words that, derivations excepted, are unconnected with one another (Webster gives synonyms), instead of offering it as a structured and patterned system, which is the way words are used in practice. The alphabetical method of organizing the entries is a primitive one, but is used for want of anything better. True, it is convenient for people accustomed to alphabetical listing and an alphabetical arrangement is better than no arrangement at all, but it has many weaknesses.

An alphabetical dictionary may display fields in the language in a tentative and irregular manner, in areas like morphology for example, with an entry such as *ease*, followed by *easy*, *easily* and so on. Contrasts that would greatly assist the reader to understand the range of the word and the exact shade of meaning are rarely given.

The entries in the *OED*, probably the most renowned of general monolingual dictionaries and admired by lexicographers throughout the world, consist of: forms (usual or typical spelling); pronunciation; grammatical designation (i.e. word class); specification (the usage in a particular technical terminology or a language variety); status (obsolete, archaic, colloquial and so on); principal earlier forms of spelling; inflexions; significations; and quotations. The significations and the quotations are most useful to the translator, as they provide an indication of the meaning and the range of application of the word (where and how it is used). But many details are left out, and can be found only in grammar books.

The standard dictionary gives only definition and quotation These do not claim to offer a full and accurate picture of the language; they merely suggest some applications of the elements of the vocabulary as isolated units. The lexicographer may try to record usage in his quotations, but he is limited to written quotations, to material recorded in texts, because in the past these alone could be verified as belonging to usage. Newspaper usage may be recorded as being more up to date, but there is no guarantee that expressions found in the press will have more than a short life. Even if the description of the vocabulary as used in living speech was comprehensive and contemporary, it would almost certainly be obsolete in some respects by the time the material had been collected, collated, edited and published. The more exhaustive the dictionary seeks to be, the more likely it is that this will happen, so the smaller versions are often better in this respect.

It is also important to note that while the lexicographer insists that he is merely describing the language as it is, this is not what the general reader thinks. To the great mass of people, dictionaries are authorities on how language should be used. Hence even dictionaries that aim purely to describe become prescriptive in practice. This fact is known to lexicographers and can scarcely fail to have some influence on their selection of forms. The general effect must be to make the dictionary even more conservative.

Modern technology promises to improve the position. Electronic aids enable the dictionary compiler to use material that has hitherto been difficult to collect. Thanks to the tape-recorder, the radio and other modern means, the spoken language can be observed and captured on paper as never before. Electronic aids also mean that a larger volume of material can be processed and dealt with, at a rate that would have seemed incredible in earlier times. Computer-produced dictionaries will have a revolutionary effect on translation. A computer is being used in a number of lexicographical enterprises at the present time. These include the Centre d'Etude du Vocabulaire Français at Besançon, which aims to produce a complete mechanized inventory of the French language. The first volume of the *Trésor de la langue française*, covering the period from 1789 to 1960, has already appeared. Use of modern data-processing methods has resulted in a total of 90 million occurrences being handled in one-twelfth of the time it took to complete and publish the *Oxford English Dictionary*. One of the more important advantages has been the possibility of including a wide range of technical terms. More work of this nature is being carried out at other centres. The translation service of the German Federal Army at Mannheim, for instance, is already using computerized technical glossaries, and to great advantage. This is undoubtedly the trend of the future. But it will be some time before the lexicographical revolution takes full effect, and until then the defective dictionaries of the pre-electronic age will continue to be used.

Some monolingual dictionaries of the conventional type do attempt to structure the vocabulary in a more rational manner. *Webster* uses synonyms; the *Martin Alonso* dictionary of modern Spanish, *Enciclopedia del Idioma*, defines by means of quotations of opposition and contrast, synonyms, paraphrases, antonyms and negations.

The standard dictionary is often dependent on secondary sources – quotations from earlier dictionaries and from literature that may be far removed from everyday speech. Where it classifies quotations it usually does so in chronological order, as in the *OED*, rather than in order of importance in contemporary usage. This is a delicate matter, since importance is difficult to judge. One factor, which electronic methods must necessarily favour, is that of frequency of occurrence. Even this, however, is by no means definite. Words that occur relatively frequently in daily use may have great weight in the vocabulary, whereas the most common words have little information value, though they are of course important functionally. Certain words (say, *constitutional*) do not occur very often but are indispensable to the language. The potential value of items is also important.

Another serious defect of the standard dictionary is its incompleteness, though this, again, is difficult to assess. How complete can a dictionary be? Obviously a dictionary including the half a million or so words of the German vocabulary must be bulky. The *OED* runs to twelve volumes without the Supplements; *Littré* has seven and *Robert* six. For general use, however, there has to be some sacrifice of completeness. Not everyone owns a dictionary, and few people consult the *OED* or *Robert*, but most people in a literate society have to consult a dictionary at some time, if only to solve a crossword puzzle. Because of the large lay public that exists today, a standard dictionary cannot be a purely scientific account of the language. It must cater for popular needs. The *OED* is for philologists, but the *Concise Oxford Dictionary* (*COD*) is for schoolboys, crossword addicts, typists, the man in the street and in the office.

All these dictionaries, large and small, comprehensive or limited, are atomistic and almost all take the form of lists of words accompanied by some information about them. This is neither a logical classification nor a scientific arrangement. The lexicographer is fettered by the habits, ideas and illusions of the people using the language, including their illusions about the authority of dictionaries. The isolation of each item, the way it is presented as unrelated to other items apart from the morphological series, and its attribution to word class, is a fundamental weakness. The dictionary thus fails to depict the language accurately in these respects.

One of the remedies that has been suggested is a conceptual dictionary. *Roget's Thesaurus*, already mentioned, has served as an inspiration to many continental lexicographers and has even served as the point of departure for researchers experimenting with machine translation. It is an excellent work of its kind, yet far from perfect. The arrangement of items in terms of ideas is a promising one, but depends on intuition. The classification reflects the views of one man; the choice of concepts is arbitrary and the structures, being conceptual, do not directly depict linguistic relations. On the other hand, it is not clear to what extent the classification depends on unconscious linguistic associations rather than conceptual ones. It is very difficult, perhaps impossible, to dissociate concepts from language forms. At the present time there does not seem to be any entirely satisfactory conceptual classification. Experience with library classification schemes and thesauri in information storage and retrieval systems yields ample evidence of the difficulties. As it stands, a thesaurus cannot replace a dictionary for the translator, although it may well be a useful supplement.

The approach adopted by a thesaurus is quite different from that adopted by a dictionary. A dictionary is based on the word, which is then related to other words and only indirectly to the external world. The definitions are marked by circularity. In *Webster's Seventh New Collegiate*, for example, the following entry occurs:

adroit\ - '*droit*\ *adj* [F.fr. *à droit* properly] 1: dexterous in the use of the hands 2: marked by shrewdness, craft or resourcefulness

in coping with difficulty or danger syn see CLEVER, DEXTEROUS
– *adroitly adv* – *adroitness n.*

The information appears in the following order: entry, pronunciation, word class, etymology, first sense, second sense. To
understand the first sense it is necessary to know what *dexterous*
means. The entry for this word follows the same pattern. The first
meaning offered is: 'skilful and competent with the hands'; the
second: 'mentally adroit and skilful: EXPERT'; the third: 'done
with dexterity – *dexterously adv dexterousness n*'. This tells us that
adroit and *dexterous* are adjectives used with the noun *hands*; they
are used to designate skill with the hands; in one meaning this skill
is used to designate mental qualities. If we look up the word *clever*
we find the same repetition. Each definition offers a slightly different range of meaning. The first meaning of *adroit* coincides with
that of *dexterous*: '*Adroit* 1 = dexterous in the use of the hands'
So *adroit* 1 = skilful and competent with the hands; mentally skilful and competent with the hands and skilful; EXPERT 3: done with
dexterity . . . Similarly, *dexterous* 2 = mentally adroit and skilful,
i.e. = mentally skilful and competent with the hands and so on.

This naturally reduces the entries to absurdity. The dictionary
contains instructions on how to use it and the majority of readers
are aware of the conventions for using dictionaries. Nevertheless
the circularity exists: *adroit* is *dexterous* is *adroit*. With patience a
series can be established of *adroit-dexterous-skilful-clever* and so on,
and the shades of meaning for each can be determined. A synonym
dictionary performs this very useful task, and *Webster's Dictionary
of Synonyms* is an invaluable resource to a translator in an emerg-
ency, just as the Duden *Vergleichendes Synonymwörterbuch* can help
out in German.

Dictionaries of the encyclopedic type, like *Webster's*, the *Petit
Larousse*, the *Duden* dictionaries, *Der Kleine Brockhaus* or the
Dictionnaire Usuel Quillet-Flammarion, use illustrations and tables to
provide denotative meanings. These are probably the most useful
general dictionaries for the technical translator.

In contrast to these, which are semasiological in approach – i.e.
they start off from the name and may arrive at the thing it designates – thesauri are onomasiological in approach – they proceed
from thing to name, from concept to word. The difference is well
explained by Roget himself in his introduction to the *Thesaurus*:

The purpose of an ordinary dictionary is simply to explain the meaning of the words: and the problem of which it professes to find the solution may be stated thus: – The word being given, to find the signification, or the idea it is intended to convey. The object aimed at in the present undertaking is exactly the converse of this: namely – the idea being given, to find the word, or words, by which that idea may be most fitly and aptly expressed. For this purpose, the words and phrases of the language are here classed, not according to their sound and orthography, but strictly according to their signification.

This statement is well worth bearing in mind, for it is an accurate account of the differences in approach and has a great deal of relevance to the needs of the technical translator, and to the specialized technical dictionary. But the *Thesaurus* itself is not completely successful.

Specialized dictionaries of slang and dialect can sometimes come to the rescue of the translator. Writers do not all have a polished command of their native tongue, and expressions slip in from various levels and registers. In addition slang, like colloquial language in general, is an important source of technical terms. In the workshop, designations are often coined *ad hoc* and hastily. Intended merely to meet the needs of the moment, they may become permanently embedded in the language. Colloquial expressions rise to more formal styles in due course. In any event they may be encountered in a technical text. A work like Küpper's *Wörterbuch der deutschen Umgangssprache* provides the answer to many a linguistic puzzle arising from the spoken language and too new for the general dictionary.

The technical monolingual dictionary can take many forms: it may be a list of words with equivalents in the same language; it may be a list with rudimentary definitions or full definitions, which may or may not have been standardized by an accepted authority. The necessary distinction between the technical vocabulary that belongs to the scientific level and the vocabulary that belongs to the workshop and other levels is not always drawn. The first tends to reflect the internationalism of science and to offer a word stock based on the classical tongues. It has clear and accurate definitions. The other is rooted in the native language and is more popular in character. It is likely to contain many old-established expressions, very rich in synonyms and transferred meanings. Ancient trades and techniques may, however, have been modernized in our own time and may have acquired new terminology or sought to improve existing expressions: in watchmaking, for example, consider the development of electronic watches, the use of watches in space exploration, the high degree of accuracy of caesium clocks; other examples are milling engineering, a craft going back to antiquity yet keeping pace today with the most advanced technology and agriculture.

If the technical dictionary is to be useful it must be up to date and accurate and must provide precise definitions. Certain parts of the technical vocabulary remain fairly constant, especially scientific terms based on Latin and Greek, but science and technology are areas of rapid growth and the greatest source of new words today. It is more difficult to keep up to date in this field than anywhere else. As the excellent *Dictionary of Technical Terms for Aerospace Use*, published under the auspices of NASA, warns: 'No set of definitions frozen in type can hope to capture the continuing change of living language.' It should be noted that many definitions in the dictionary are based on those included in the published standards of technical societies and standards organizations; but it is not a terminological dictionary in which all definitions are standardized. Its compilers point out that the definitions should not be used as standards, and that the rigorous requirements for standards have not been made. It lies, in short, somewhere between a dictionary of standardized terms, a terminological dictionary and a general monolingual dictionary. It is a specialized dictionary and has the characteristics of general dictionaries – the entries carry meanings that are designative as well as denotative. It belongs in part to scientific language and in part to workshop language. This is true of many technical dictionaries but it is especially noticeable in the building industry, and in medicine and mining, while electronics, chemistry and the newer sciences have more standardized terms.

The bilingual dictionary has a particular importance for the translator, but it is also a very dangerous tool. In general when a translator needs to resort to a dictionary to find an equivalent he will do better to consult a good monolingual dictionary in the SL and, if necessary, one in the TL as well. The bilingual dictionary appears to be a short cut and to save time, but only a perfect bilingual dictionary can really do this, and no bilingual dictionary is perfect. There are some very good ones and quite a large number of middling, indifferent and bad ones. Obsolescence is more of a threat here than with monolingual dictionaries. Bilingual general dictionaries differ from monolingual ones in offering translation equivalents instead of definitions. They may also provide quotations and examples of usage. These are often fixed modulations. The equivalents may be one-to-one word correspondences, even one-to-one morphemic correspondences, or single words replacing word groups or compounds, and the reverse, established collocations (e.g. clichés and stereotyped phrases) replacing established collocations or single words or word groups. The dictionary may work in one direction only, from an SL to a TL, or it may work in two directions, SL to TL and TL to SL, in two sections.

Definitions that are wholly appropriate and desirable in a monolingual dictionary are unnecessary or misleading in a bilingual one. The *Cambridge Italian–English Dictionary*, which is one of the better examples of its type, illustrates this with an example from one of its predecessors. There the word *maglia* was paraphrased in English as 'closely knit body-covering worn by dancers to simulate nudity'. The word required was *tights*. As with translation in general, the guiding principle is to match the smallest possible units in the two languages. A long definition in a bilingual dictionary is merely frustrating to the reader who wants a word. At best it is better than nothing, but that is not saying very much. A monolingual Italian dictionary of any merit will be more helpful than this, since it will supply, in addition to definitions, synonyms or near-synonyms. In this way the designative range of the Italian word will be better conveyed. A single equivalent is not satisfactory, because it is bound to be a selection from a series of possible alternatives, and the selection is made according to the arbitrary judgement of the compiler. Neither the definition nor the single word equivalent meets the needs of translation. Dictionaries that give only the latter are to be avoided like the plague, as they are much more misleading than the first group. The ideal, as expressed by the authors of the *Cambridge Italian–English Dictionary*, is 'the mapping out of the whole area of signification of Italian words by means of a series of equivalents, which not only serve to make the meanings and usage clear, but also provide, or failing that, suggest, the exact or most appropriate translation for the context in which the user has seen or heard a word or phrase'.

These requirements will of course make the dictionary bulky and expensive, but anything less will be a false economy. The *Muret-Sanders* German–English dictionary complies with these requirements, but it is dated, and a new version is being published by Langenscheidt. There are other good German–English bilingual dictionaries, of course, and chapter 17 offers advice on sources of information. For French–English the most useful dictionary is probably *Harrap*, edited by J. E. Mansion. These two are indispensable for anybody seeking to translate from the languages mentioned, but a technical translator should never depend on a general bilingual dictionary alone, no matter how excellent.

At its best a bilingual dictionary necessarily shares the main limitations of a monolingual one. It has the same alphabetic and atomistic classifications, the same tendency to obsolescence – indeed this last factor is even more marked because two languages are concerned instead of one. Moreover, bilingual dictionaries tend to furnish standardized translations that do not correspond to the full lexical ranges in the two languages and may therefore be incorrect because of shifts of meaning in both languages.

The role of the bilingual dictionary, and to some extent that of the monolingual dictionary, is to give approximations rather than true equivalents. There may be a one-to-one correspondence in a particular context, but the correspondence may just as easily be many-to-one or one-to-many. A bilingual dictionary is like a direction post that tells us to proceed north but does not tell us where to turn off or to stop. In the words of the *Cambridge Italian–English Dictionary*: 'The "telegraphic" nature of dictionary-language is such that it depends for comprehension upon the maximum possible co-operation of the user. In this connection, also, the principle of orientation is influential. In the selection of near synonyms, in the distinction drawn between styles and categories, in the association of ideas serving to determine groupings, reliance has been placed on the English-speaking user's understanding of his own language.'

The bilingual dictionary should in fact be used only as a last resort. It is not the first aid to be sought. The first dictionary that should be used is a terminological one, if that is available. Next should come a technical dictionary dealing specifically with the subject field. If that is not available, a general technical dictionary can be used. Failing these, and if the problem is one of general vocabulary, the resources of the monolingual dictionary should be explored. This is more likely than the bilingual dictionary to lead to the concept underlying the lexical item and its associations. But if none of these is available, then only a good bilingual dictionary should be used. Bad ones are dangerous.

What makes a good bilingual dictionary? The general features are outlined above. It should provide correct translation equivalents – the difficulties involved should by now be clear; it should provide as wide a range of application as possible for each item – the range will never be wide enough; it should provide full grammatical information – word class, inflectional and derivational forms, syntactic restrictions and applications; it should give the level of usage of the equivalents provided.

The most relevant dictionary here is the technical dictionary. It is necessary to distinguish between a terminological dictionary and a technical dictionary such as the aerospace dictionary already mentioned. A terminological dictionary is a dictionary listing standardized terms and their definitions.

Before describing the terminological dictionary I shall recapitulate certain characteristics of the term and add to the information already given. The requirements for a properly designed term are: nonambiguity, simplicity and euphony. A standardized term may be used with only one meaning. Definition of a term means abstracting from it all characteristics, or all portions of its designative range, except what is strictly relevant to the technical field. It should express only one concept, the concept being defined before the term is created. Definition in a general dictionary is 'lexical'. It is mainly a description or explanation of the way in which a word is used at the time of compilation. In contrast to this, a 'stipulative definition' is required.[1] This implies deliberately assigning an object or concept to a word, or assigning a word to a concept or object. A terminological dictionary thus consists of standardized terms and their stipulative definitions or assignments to standardized concepts, one term to a concept.

A model for such a dictionary is the *DIN-Begriffslexikon, Benennungen und Definitionen aus den deutschen Normen* ('DIN Conceptual Dictionary, Designations and Definitions from the German Standards'). This contains standardized definitions with DIN numbers and cross-references to other definitions in the book. The nature of the definitions is recognized in a prefatory note explaining that standards are the product of agreement to act in accordance with them.

The technical dictionary, as distinct from the terminological one, does not contain only standardized terms. This is justifiable on the principle, well recognized by those concerned with technical language, that words that have become established by long usage should not be removed from a vocabulary unless there are strong and obvious reasons for doing so. A further principle is that theoretical criteria should not be allowed to take precedence over practical ones. This is very relevant to those spheres where workshop language is prevalent. In this instance convenience is to be preferred to accuracy.

Technical dictionaries may be monolingual general dictionaries, monolingual specialized dictionaries, bilingual general dictionaries, bilingual specialized dictionaries and multilingual general or specialized dictionaries.

The monolingual general technical dictionary covers the entire field of science and/or technology. It is more likely to be useful to the layman than to the specialist, but it can be valuable for quick reference. *Chambers Technical Dictionary* is concerned with a wide range of science and technology, as is the *Elementary Scientific and Technical Dictionary* of Flood and West.

Monolingual specialized dictionaries are confined to a single discipline and range from works that are almost encyclopedic in scope to lists of words with short definitions. The task of compiling them is more difficult in the 'growth' subjects than in the old-established ones.

This draws attention to a basic dilemma of the compiler of technical dictionaries. Although we have noted it in the general dictionary as well, it has special significance here. The technical dictionary is most needed precisely where there is uncertainty about a term because it is still new and its permanence is not yet assured. But the same problem exists for the dictionary compiler, so it may be a case of the blind leading the blind. In practice the compiler will be tempted to wait until the new expression is established or has been eliminated. But by the time he has taken his decision the expression has become established or has fallen out of use and there is no need to look it up at all. It is natural that the lexicographer should be cautious, because a technical dictionary is a business venture. However, the rate of obsolescence in technical vocabulary is very high, higher and faster than in the general language. In his introduction the author of the *Penguin Dictionary of Electronics* (1962) observes that of the 5000 or so entries about a third did not exist before 1950, 'yet many of the new words are already part of our common language'. The bilingual technical dictionary suffers from this problem to a lesser extent. The general bilingual dictionaries are perhaps worst off in this respect. They need to be comprehensive, and comprehensiveness is difficult to assess. A large number of entries may include expressions adequately covered by ordinary dictionaries. Comprehensiveness may also mean delay and dating. Lack of comprehensiveness may mean that the translator consults the dictionary in vain and wastes time. It may also mean that only one equivalent is given for an expression that occurs in different subject fields, with different definitions in each. Among the best general bilingual technical dictionaries are the De Vries *German–English Science Dictionary* and his *French–English Science Dictionary*. Kettridge's *French and English Dictionary of Technical Terms and Phrases* is also useful. For Spanish, S.A. Castilla's *Spanish and English Technical Dictionary* may be mentioned.

A superb work (as indeed all his publications are), though not conforming to the pattern of a conventional dictionary, is Henry C. Freeman's *Technisches Englisch*. This also contains diagrams showing the 'family' relationships of terms in engineering. Here is an example:

Spezialwalzwerke
Special mills

Röhrenwalzwerk	Radreifenwalzwerk	Radschreibenwalzwerk
Tube mill	Tire mill	Wheel mill

Freeman's explanations by means of diagrams, quotations and definitions of the ranges of applications of expressions such as *Leistung, Betrieb, Diagramm, Bearbeitung, Messen* or *Messung* are exemplary. His books cannot be too highly recommended to those attempting to translate material on mechanical engineering from German to English. Unfortunately he works only in this field of technical lexicography.

Multilingual dictionaries are found more in technical fields than anywhere else. They vary in quality. Again general and specialized versions are available. Among the general type, Schlomann's *Technological Dictionary* in three volumes, dealing with English, French and German, may be mentioned. These dictionaries either simply list equivalents or provide definitions for each entry, sometimes in only one language. Those with definitions are to be preferred and among them those with a separate definition in each language for each item. Very often these dictionaries adopt a system of numbering for the entries that is useful for reference purposes. Some of these are bad, some mediocre and some good. As with bilingual technical dictionaries, they should not contain words that are found in general dictionaries, or indeed in other technical dictionaries. Nevertheless, this is too often done to pad out the dictionary and give an impression of comprehensiveness.

The model for these dictionaries is a terminological one, *The Machine Tool: An Interlingual Dictionary of Basic Concepts*, produced on an international scale under the aegis of the United Nations, the International Standards Organization, the International Federation for Documentation and five national governments. The guiding spirits are well-known names in the field of terminological studies, including Dr Eugen Wüster (ISO and FID) and Dr J.E. Holmstrom (Unesco). The dictionary is published in an English–French master volume and has a German supplement, *Grundbegriffe bei Werkzeugmaschinen*. Further supplements for other languages are planned. This dictionary is not only marked by accuracy and completeness. It attempts to make full use of the work on standardization in the leading industrial nations and aims at a high degree of precision in the presentation of terms and concepts. Every concept is carefully defined and is often accompanied by illustrations. It follows the recommendations of the ISO on dictionary layout. The introduction to the dictionary is in itself a valuable contribution to technical lexicography and terminological studies. The entries are in alphabetical order and are accompanied by numbers that provide easy access to equivalents in other languages.

In its 1957 report on scientific and technical translating Unesco made a number of recommendations for specialized dictionaries.[2] It suggested that:

a) each organization should concentrate on its own special terminology in compiling dictionaries
b) terms and phrases should be differentiated into three categories – expressions likely to be encountered in the literature but not recommended; recommended expressions; standard expressions referred to officially adopted definitions

c) dictionaries for more than two or three languages should be divided into a basic volume giving the equivalent terms for these languages together with illustrations, definitions or contests, and a supplementary volume for each additional language.

The desirable qualities for a technical dictionary can be summed up as follows: no terms that occur in a general dictionary should be included; terms should not be included if the constituent parts clearly show the meaning of the whole; each dictionary should cover only a limited and specialized field; no general terms common to science and technology as a whole should be included. For the translator's needs, the dictionary need be in only one direction, from SL to TL.

CHAPTER 16

Machine translation

𝕊𝕊𝕊𝕊𝕊𝕊

THE relatively brief history of machine translation (MT) has been marked by a spate of highly optimistic claims and forecasts countered by disbelief in any prospect of superseding the human translator. The periodic announcements of imminent 'break-throughs' have subsided and the first flush of enthusiasm appears to have died away. There is still an optimistic school and still a pessimistic one, but between them a third 'moderate' view has emerged.

The problem of MT is essentially that of converting the phenomena of language into a form that can be processed by an electronic digital computer, that is, into a numerical form. It is now clear that earlier experimenters did not appreciate the complexity of these phenomena. They saw language as a code, in the sense that it was a limited system of signs, and thought to translate was to replace each sign in the original system by an equivalent sign in the new system. Warren Weaver, who is considered to be one of the initiators of the postwar MT experiments, regarded the task as akin to that of deciphering enemy codes.[1]

Shortly after the war Dr A.D. Booth, of Birkbeck College, London, approached Warren Weaver of the Rockefeller Foundation and suggested that computers should be used in translating. A memorandum by Weaver is one of the landmarks in the history of the MT project. He wrote that the multiplicity of languages was a serious deterrent to international understanding and foresaw the use of high-capacity electronic computers in helping to solve the problem.[2] The work soon spread from the United States and Britain to Russia and research was eventually carried out in a large number of countries. Articles appeared in newspapers and periodicals, their authors virtually proclaiming MT as an accomplished fact. Even as late as 1959 Booth was envisaging a rosy picture: 'Not only does it now appear possible that translations of good quality can be made from both scientific and literary texts, but also that some of the recent developments in machine technology will make it possible to read directly from the printed or typewritten page.'[3]

This is an exciting prospect, but it is clear by now that many bridges have to be crossed before it can be achieved – if it can be achieved at all. Difficulties began to arise almost immediately, and even the most optimistic researchers began to despair. Today the sights are set on less ambitious targets than the initial ones of fully automated translation. Computers are looked upon as an aid to the human translator, rather than as a substitute. This is not to say that some researchers do not still hope to attain fully automated translation.

The disciplines whose combined resources are involved in the MT project are linguistics, mathematics and engineering. Linguistics seeks to provide a structural description of language that can be expressed in the terms of information theory, and which can in turn be used by data-processing computers. For this a logical basis must be found in language. The first MT experimenters were scientists and engineers, not linguists. They looked upon the problem as purely and simply one of engineering. They did not view the reduction of language to numerical data as particularly difficult in principle, and believed that the facts could all be stored in a machine of sufficient capacity. Language in these terms is a finite system obeying clearly defined laws and organized logically, and thus susceptible to quantitative analysis. Some elements of language do lend themselves to binary representation (marked–unmarked† items like plural *v.* singular in many languages).

In any event the MT researchers began with word-for-word analysis, found this inadequate and proceeded, in 1952, to deal with sentences. This was of course an implicit admission that the problem was much more complicated than they had at first realized. The next big landmark occurred in 1954, when Leon Dostert and Paul Garvin gave a famous demonstration on an IBM computer at Georgetown University. This involved programming 250 words and six simple syntactic rules and has passed into history as the 'Georgetown-IBM demonstration'. In the same year

---

† Unmarked terms can be regarded as the normal form and marked terms as more complex and specialized. In English the plural is said to be marked by final *s*, whereas the singular is unmarked.

the journal *Mechanical Translation* was launched by Victor Yngve at Massachusetts Institute of Technology. The year 1956 saw the first international conference, organized by MIT, in which the Soviet Union took part. Further conferences of note were the International Congress of Linguistics, held in Oslo in 1957, the Los Angeles conference of 1960 and the 1961 conference in Teddington.

At this time it was estimated that there were 68 research groups busy on the problem, 32 of these in the United States alone. The rest were distributed as follows: Russia 8, Great Britain 8, France 4, 2 each in Italy, Japan and Yugoslavia, and 1 each in Belgium, Czechoslovakia, West Germany, Israel, Mexico, Red China, Rumania, Sweden, plus 1 each organized by Unesco and Euratom.[4]

There are a number of parallels between MT and human translation. This is to be expected, as both are forms of translation from natural languages. But in MT instructions have to be formulated that are intelligible to a computer, and a computer, however complex and marvellous, is still a machine without initiative or imagination. It has no creativity, or any of the other difficult-to-define qualities that a human translator must possess. These instructions must not leave anything to chance, so every detail must be determined with the utmost precision. It has often been said that a computer is an idiot, and many examples of its imbecility (probably apocryphal) have been given. 'The ghost is a volunteer, but the meat is tender', an alleged computer translation of *der Geist ist willig, aber das Fleisch ist schwach*, underlines the fundamental problems inherent in MT. The machine's one advantage, an inestimable one, is its vast storage capacity and its speed. The problems involved must therefore be expected to include those of human translation, together with others peculiar to itself. The engineering problems are soluble – but the linguistic and other problems, such as those described in earlier chapters, are elusive.

The history of MT largely recapitulates the fallacies and misconceptions discussed earlier. As we have seen, it started with the simplest approach – the replacement of a sign (equated with a word) with another sign. This necessitated the compilation and storage of a bilingual dictionary. Each word in the SL had to be looked up in this dictionary and an equivalent in the TL had to be selected. At the end of the sentence the equivalents were rearranged in accordance with the word order customary in the TL.

The research workers soon realized that a series of word-for-word equivalents in the SL sequence merely led to a kind of pidgin. It has been claimed that even this can be useful to a scientist who knows the subject, but in practice the deciphering of this gibberish is frustrating and time-wasting. Rearrangement of the words in a more acceptable order before feeding them into the computer gave better results, but this meant that the translation was in part the work of a human editor. When we remember that much key-punching is also necessary to record the dictionary, it is understandable that MT must be much more expensive than human translation. Not only must the editor rearrange the words, he also has to select the correct equivalents. Since one-to-one correspondences do not occur with any degree of regularity between two vocabularies, a computer cannot cope with the assignment of equivalents.

Two schools of thought now emerged. The dispute centred on what Bar-Hillel, a leading researcher into the problem, called 'fully automatic high-quality translation' (FAHQT).[5] He was not sanguine about the possibility of achieving this. At one extreme were those who were impressed by the complexity of language and were thus pessimistic about the possibility of processing it in a computer. At the other were those who were impressed by the capabilities of the computer and were inclined to minimize the language problem. This division persists to this day. The first group include the 'perfectionists' and the second the adherents of the 'brute force' approach.[6] The latter believe that with a sufficiently large memory, machine translation can be carried out without a complex set of instructions – either by means of a very large dictionary or by means of a large dictionary plus a table of grammatical rules (stored in the computer memory). The 'perfectionists' believe that without a complete theoretical knowledge of the SL and TL and a perfect understanding of the process of translation, the task is impossible. Garvin and his associates agree that extensive knowledge of languages is needed, but hold that the type of knowledge needed for designing MT systems is primarily empirical knowledge, 'and above all problem-solving engineering know-how'.

At present the situation in MT is somewhat static. Work is proceeding at various centres but those concerned are more cautious in their hopes and predictions than the earlier experimenters.

Meanwhile much valuable experience has been gained in the way of syntactic analysis and in the compilation of dictionaries.

All MT systems include a machine dictionary and an algorithm An algorithm is a set of instructions about the steps to be taken by the computer. The dictionary contains SL words and TL equivalents, together with a set of codes to instruct the algorithmic portion to consult the dictionary. The MT dictionary encodes morphological information about parts of speech and inflectional characteristics, and possible syntactic information as well. It also contains semantic codes. The first grammatical code carries out automatic parsing while the second resolves semantic ambiguity.

The problem of MT is a problem not so much of the machine as of translation. As in ordinary translation, the principal difficulties arise from constant innovations in language, changes in meaning, polysemy and ambiguity, and from the cultural background implicit in a text. The adequacy problem is perhaps even more urgent here, since if MT is to become a practical method of translating it must produce usable translations cheaply as well as speedily. Speed is of course its one great advantage over the human translator, but it falls down on quality and cost.

It has been suggested that even at its present stage a computer can produce translations that convey enough information to meet the needs of a scientist. This view is based on the extensive use of standardized terminology in technical writing, the internationality of research and the simple syntactic structures prevalent in this type of text. The result is in effect a 'rough' translation as described in the previous chapter – one that may enable the user to decide on whether or not the material concerns him, and on whether or not it is worth translating. But even in the case of technical texts this function is of limited value.

Essentially MT depends on the possibility of codifying language in accordance with strict laws and constant relationships. But natural language is unlike the artificial computer languages, such as Algol and Fortran, which have been specially designed for a computer.

A computer is a machine designed to carry out instructions that must be converted into electronic impulses. A computer language is an artificial code designed to enable the programmer to express algorithms or natural languages in a series of processes that can be directly converted into machine language. Computer

languages allow the programmer to write his programme in something close to a natural language or in mathematical notation and
to allow the computer to convert the translation into machine
code. The machine often has a special program incorporated into
it that transforms computer language into electrical code. Natural
language is not suitable as a direct input because of its lack of
logical construction and because it does not present its information
on a step-by-step basis. More important, many processes in natural
language are not explicit or formally expressed.

Each computer language is designed to fulfil specific purposes
and with specific advantages in mind. The name given to it is
generally an acronym specifying these purposes. Algol is an
'algo[rithmic] l[anguage]'; Fortran is 'for[mula] tran[slation]'.
Both Algol and Fortran are scientific languages, whereas Cobol
is a 'co[mmon] b[usiness] o[riented] l[anguage]'.

Fortran is used mainly for processing figures, for example at
Euratom. It was devised by IBM, who have also developed a newer
language called PL/I. Algol 60 is used particularly in the English-
speaking world, in Scandinavia and in The Netherlands, in
mathematical linguistics. Comit was developed at MIT for machine
translation and is little used outside Massachusetts. Cobol is used
a great deal in the Romance-language areas. It is related to the
better-known Algol.

All these are mathematical languages, deliberately created and
thoroughly standardized instruments. No matter how complex
they are, and they are constantly being developed and improved,
they are fundamentally different from natural languages. To a
limited extent language can be codified, but statistical methods
have not proved effective so far.

Bar-Hillel has warned against looking at quantities, since one
is caught up in the fallacy that if, say, 80 per cent of a text can
be translated, the work on the remaining 20 per cent will take
only a quarter of the time: 'The remaining 20% will require not
one quarter of the effort spent for the first 80%, but many, many
times this effort, with a few per cent remaining beyond the reach
of every conceivable effort.'[7] His argument becomes convincing
if we remember that the remaining unsolved problems are precisely the very difficult ones that cannot be reduced to mathematical formulations or simple correspondences, the problems of
ambiguity. How is the machine to know when to choose *meat*

rather than *flesh* for *Fleisch*? How is it to cope with metaphor and transferred meanings? Language is full of inconsistency, polysemy, idiom, stylistic idiosyncracies, coinages, illogical compounds, variations of levels; and even technical terminology leaves a good deal to be desired in this respect. The human translator can have a hard time steering his way through these stormy seas, but he has special qualities that make it reasonably probable that he will reach his destination. The electronic computer does not possess these qualities (which are described below). Syntactic ambiguity is very troublesome for the machine but can often be resolved by human translators without too much difficulty. Semantic ambiguities are infinitely more difficult, and even human translators cannot cope with all of them. This does not imply that progress will never be made. Nor does it mean that the work on MT is valueless. On the contrary, it is of the highest significance. It has been instrumental in the development of a new understanding of the workings of language, thanks to the necessity for making everything explicit for the computer. The work done on computerized dictionaries and glossaries is of great importance. Searching for terminological and lexical equivalents represents a large proportion of human translators' time. Studies by the Federal Armed Forces Translation Agency at Mannheim in Germany found that a translator working with conventional aids requires between 50 and 86 per cent more time than a translator working with a glossary specially compiled for the text in question.[8] Working with such a glossary, translators also made one-third fewer errors.

In the German army's system, the translator reads the text and underlines the SL words for which he requires the TL equivalent. The text is given to a keypunch operator, who prepares cards for the expressions underlined and makes the necessary morphological changes (e.g. eliminating inflectional suffixes). The information is then fed into the computer from which such glossaries can be produced by the computer in about ten minutes. This system has been in use since 1965. It works best with experienced translators.

The work on MT has also been of great value for language data-processing in general – data storage and retrieval, thesaurus construction, key-word systems and so on.

CHAPTER 17

Extralinguistic sources

𝕊𝕊𝕊𝕊𝕊𝕊

T H I S chapter is concerned with what happens when the translator
has to go beyond the grammar book and the dictionary to arrive
at an understanding of the SL text. It surveys the sources of background information, and at the same time describes the different
types of text a technical translator may have to deal with.

I have assumed that the translator has access to a library, preferably a specialist technical library. If not, unless he is a specialist
in the subject himself, he will be unable to solve certain problems,
or at any rate to solve them speedily. An ideal situation occurs
when a translation service is associated with a technical library.
This means that the translating process is part of a general communication process – a foreign-language information activity, which
is exactly how it should be seen.

Good documentation is essential for the technical translator. But
he should be able to consult other sources than documents, such
as subject specialists. A person who is already working in the field
will normally have more up-to-the-minute information than a
printed document. But documents can still be of vital importance,
and one cannot be continually taking up the time of a subject
specialist.

The first step to take when a translation is requested is to make
sure that one has not already been made elsewhere. This may seem
obvious, but it is so obvious that it is sometimes neglected. A
number of records of published and unpublished translations are
available. The Association of Special Libraries and Information
Bureaux (Aslib) in Britain publishes a *Commonwealth Index of Unpublished Scientific Translations* in cooperation with the Canadian
National Research Council, the Commonwealth Scientific and
Industrial Research Organization of Australia, the New Zealand
Department of Scientific and Industrial Research, the South African Council for Scientific and Industrial Research and the Indian

Ministry of Natural Resources and Scientific Research. This is a register of available translations and Aslib (or any of the co-operating bodies) will put an inquirer in touch with the organization that holds copies of the translation required. The National Lending Library for Science and Technology in Boston has a collection of more than a hundred thousand items covering the whole field of science and technology. Any items can be borrowed or consulted and photocopies will be supplied. The European Translations Centre at the University of Delft in the Netherlands is a clearing house for translations produced by some national translation units. It will supply photocopies and will refer inquirers to those who hold the translations required. It also publishes a quarterly *World Index of Scientific Translations*. Other European translation pools are to be found at the Zentralstelle für Wissenschaftliche Literatur in Berlin and the Centre National de la Recherche Scientifique in Paris. The most comprehensive index of scientific translations is the fortnightly *Technical Translations* of the Clearing House for Federal Scientific and Technical Information in Springfield, Virginia. The Special Libraries Association collection in the United States is housed in the John Crerar Library at the University of Chicago. It publishes new accessions in the *Translations register-index*.

'Cover-to-cover' translations are renderings of complete journals from a foreign language into English done on a regular basis. There are some two hundred of these, covering a wide range of subjects. There is some controversy about their value because of the inevitable waste entailed when irrelevant and unimportant material is translated, and because of the expense and the delays entailed in translating. The NLL in Britain produces a regular *List of Periodicals from USSR and 'Cover-to-cover' Translations*. Information on translated books is contained in Unesco's annual *Index Translationum*.

If a translation has definitely not been made elsewhere, the translator, after studying the text, prepares his documentation. This consists of the following:

Previous translations available on the same topic
The work to be translated may be a contract specification, a research paper, a report of a test, a trade catalogue, publicity material, an instruction, a patent application, a standard, a letter and so on.

Any document of this nature may serve as a source of information for another task.

Many sources of information are 'camouflaged' in the sense that they were not produced in order to supply the data that can be found in them. Sales brochures may be a source of up-to-date terminology, as may catalogues. The translator must not be too literal-minded in the matter and should exercise his ingenuity.

Encyclopedias

These, like dictionaries, may be general or specialized.

General, humanistic encyclopedias like the *Britannica*, *Chamber's*, *Brockhaus* or *Larousse* are of only occasional value as they are not specific enough and do not contain sufficient detail for the needs of technical translation.

Technical encyclopedias may cover the entire field of technology and science, as does the fifteen-volume *McGraw-Hill Encyclopedia of Science and Technology*. This is more likely to be useful than the humanistic encyclopedias, but the works of greatest value are specialized technical publications such as N. W. Kay, *The Modern Building Encyclopedia* (1955) or the *Encyclopedia of Polymer Science and Technology* (1967). The way in which the second encyclopedia was used to solve a terminological problem illustrates my point. The problem entailed distinguishing between *feuerhemmende Bauteile* and *feuerbeständige Bauteile*. The encyclopedia has an entry of nine pages on fire retardancy, with a discussion of terminology extending over more than one page, and including the following remarks: 'The field of fire-retardant polymers has been plagued for years by a plethora of terms without clearly defined meanings. Such terms as fireproof, flame retardance, fire resistance, flame resistance, and noncombustible have been used indiscriminately and often interchangeably.' The article continues: 'Such terms as fireproof, noncombustible, and fire resistant are somewhat ambiguous, unless qualified by a description of the fire conditions of a recognized test method.' The translator then consulted Meyer's *Lexikon der Technik und der exakten Naturwissenschaften* (1970), which offered the following: 'Feuerhemmende: Eigenschaft von Stoffen, die bei Normbrandversuch (30 min) nicht entflammen und Feuerdurchgang verhindern (unter rechnerzulässiger Belastung)' ('Fire-resisting property of substances that do not ignite under standard fire-testing (30 min) and inhibit the progress of fire . . .').

The polymer encyclopedia differentiates between intumescent fire-retardant coatings and non-intumescent ones, the former insulating the material for a substantial period from the high temperature of fire. The translator could thus translate *feuerhemmende Bauteile* as *intumescent building elements* and *feuerbeständige Bauteile* as *fire-retardant building elements*.

Handbooks

These usually deal with problems of fact within a subject field and consist of a single volume. An example is L. E. C. Hughes (ed.) *Electronic Engineer's Reference Book*. Such handbooks are primarily reference sources for engineers in industry engaged in development work, and are eminently practical in outlook.

Monographs, treatises, textbooks

Treatises are virtually encyclopedic in their scope, but are more comprehensive and exhaustive than encyclopedias. Classic works of this type in German are the Beilstein *Handbuch der organischen Chemie* and Gmelin's *Handbuch der anorganischen Chemie*.

Monographs tend to deal with a narrower field and may contain previously unpublished material. Textbooks are intended for teaching and are more limited in their approach than monographs or treatises. Their most useful function is that they facilitate a quick survey of the field and help the translator to decide where information is located.

Directories, yearbooks

These help with the translation of names of firms and organizations and can also provide terminology – names of equipment, including trade names.

Periodicals

A research paper will usually contain references to previously published work in the field. Consultation of these can provide assistance with difficult terminology but it is likely to be time-consuming. Journals include those reporting original research, those surveying, commenting on and evaluating the research in the research journals, technical journals associated with industry, trade journals and popular publications (such as *Popular Mechanics*).

Abstracting and indexing journals

These are many and varied and are useful for locating sources of information. Many periodicals contain abstracts of articles in their own issues and published elsewhere. Journals devoted entirely to abstracts include *Biological Abstracts* and *Chemical Abstracts,* while indexing periodicals include *Science Citation Index* and *Applied Science and Technology.* The National Federation of Science Abstracting and Indexing Services in Washington publishes *A Guide to the World's Abstracting and Indexing Services.*

Standards and trade literature

Standard specifications are a particularly valuable reference source, since they contain authoritative terminology. Both British and overseas standards are available at the British Standards Institution (BSI). Some BSI publications (and those of similar institutions in other countries) are glossaries. An example is British Standard 3138: 1959, entitled *Glossary of Terms in Works Study.* The United States of America Standards Institute approves standards produced by other bodies and publishes an annual *Catalog of Standards.* The publications of the American Society for Testing and Materials are of special interest. The National Bureau of Standards has an information service on standards prepared by American organizations. The German Bureau of Standards (Deutscher Normenausschuss) publishes Deutsche Industrienormen (DIN); some are also issued in English versions. The BSI publishes a monthly listing of standards from all over the world. *Overseas and Commonwealth Standards.*

The International Organization for Standardization (ISO), a United Nations group, makes recommendations for members. This body plays a leading role in the task of creating standardized international technical terminology.

The mass of trade literature, which appears in virtually every conceivable shape and size, represents an excellent source of information, though this is often 'disguised', as mentioned above. Language levels vary from colloquial to scientific, including workshop and sales systems.

Literature of this nature is either intentionally or unintentionally informative. It may take the form of lists of products, descriptions of substances or equipment and their uses, or even lavish books. A favourite form of publication is the loose-leaf volume.

Even diaries issued for professional and technical customers by commercial organizations can prove invaluable. One of the great advantages of this type of material, apart from its compact form, is that the information is up to date and the terminology is that currently used in the trade. Tables and measurements included are sometimes helpful, while illustrations make it possible to identify products and parts of equipment. Much of this material is elaborate and it includes service manuals and house journals that are often of the highest quality as far as both presentation and contents are concerned. A good deal of original research is published in them.

Theses and dissertations

These can usually be borrowed or photocopies can be provided. Aslib publishes an annual *Index to theses accepted for higher degrees in the Universities of Great Britain and Ireland*, and University Microfilms in the United States issues monthly *Dissertation Abstracts*.

Additional sources of information

These include maps and atlases, which are important for place names and range from major works, such as Bartholomew, Oxford and *The Times* atlases, or the *Goldman Grosser Weltatlas*, to smaller and handier publications.

CHAPTER 18

Competence and performance

𝔊𝔊𝔊𝔊𝔊𝔊

THE qualities required of a technical translator are a very rare combination: he or she must have a broad general knowledge in addition to his language abilities; he should also have a technical background; he must be intelligent; and he should have the ability to express himself clearly in his native tongue. In due course he will have experience to support these qualities, all of which are needed if he is to function as what Georges Mounin calls 'an invisible man', since his success is measured by the extent to which his work is not noticeable.

Any translator must obviously have a knowledge of a foreign language in addition to his own. A translator should normally translate into his mother tongue only. Many translators have a knowledge of two foreign languages, though to achieve competence in more than two is very rare. Competence in many languages is a variable quality. It means that the individual has a current competence in one or two languages besides his own and a latent competence in others – he has learnt them in the past, but he would have to work up his knowledge in any one of them in order to translate from that language. It is one thing to be able to order a meal, ask a direction, carry on a simple conversation and quite another to have complex discussions. Active knowledge of a language means being able to use it for speaking, which demands different orders of competence, depending on circumstances, and for writing, which demands its own degrees of competence. Passive knowledge means being able to read the language and/or to follow it in speech. The same qualifications about degrees of competence apply here. In principle a technical translator can get by with a good passive knowledge of the SL, but an active knowledge is still desirable, if only because of the additional understanding it gives. A good passive knowledge is more than a minimal reading knowledge. The translator should also have a good background in the literature, culture and history of the country from whose technical texts he is translating. It is possible to have a passive knowledge of several languages.

The mental processes that underlie translation are not yet clearly understood. It has been suggested that the mind stores concepts rather than words, although the process of memory also involves words.[1] Words used in conversation or discussion act as a key to the ideas expressed by the speaker. Translation can therefore be understood as a process in which words are converted into concepts and the concepts are restated in words. Two important activities occur – understanding and interpreting an utterance, which is itself a complex phenomenon; and transforming this understanding into the verbal symbols of the TL, i.e. the process of expressing the utterance in the TL. Studies of bilingualism tend to confirm this theory. An individual who is truly bilingual 'thinks' in each language; he does not merely substitute words. It has also been observed that it is easier to understand a message than to create a message in either of the two languages, including the creation in language B of something originally expressed in language A. This corroborates the observation that an active knowledge of a language is more difficult to acquire and maintain than a passive one. True polyglots, people with the ability to read, speak and reason fluently in a number of languages, are rare. Such highly gifted people do exist, but many apparent polyglots display only superficial skills. There is a vast difference between the performances of, say, a waiter and a professional translator.

Some form of technical background, ideally a university degree in a science subject or in engineering, is a great asset to the technical translator, who should have specialized knowledge in his subject field. But as with language knowledge, the potential range of the average person is limited. A translator claiming to be an expert in several disciplines should be regarded with suspicion.

The great advantage of team-work in translation – working in a private or institutional translation unit – is that it enables a large range of languages and subjects to be dealt with adequately. It also allows for a greater degree of specialization in translating proper, as the functions of documentation and establishment of terminology, translating, revising and editing can be shared out as well as permitting a pooling of knowledge.

These desiderata are ideals and they cannot always be realized in practice. Versatility may be forced on the translator. This will make his work more interesting, but it is not the most efficient way to work.

In addition to the qualities enumerated, the translator must of course be accurate. A touch of pedantry is even desirable, particularly for revision and editing, though it must be accompanied by imagination. Translation is full of such paradoxes. The translator must, for instance, be a fairly rapid worker, but he should be strong-minded enough to resist demands for speed at all costs.

A few words may be said in passing on technical interpreters. In many ways their activity is even more demanding. The interpreter must have a sharp mind and fast reactions; he must also have physical and mental stamina. Interpreting is either consecutive (the speaker says a few words, the interpreter then renders it, the speaker continues) or simultaneous – the more widespread practice today. Interpreters can improvise more freely than translators, who are bound to written texts. Properly qualified interpreters are a rarer breed than translators, and there are said to be less than 1250 of them in the world.

In the past training for translators has largely consisted of on-the-job training. Usually language graduates have entered translation offices and picked up the technical background and translating experience as they went along. Equally, science graduates with a knowledge and a flair for languages have become translators. But it is harder to find technically trained people willing to become translators than individuals with an education in an arts subject.

A number of centres have offered translation courses for many years. A number of institutes have been established at various universities and are members of the Conférence Internationale Permanente de Directeurs Universitaires pour la Formation de Traducteurs et d'Interprètes (CIUTI). The universities are those of Antwerp, Geneva, Georgetown, Heidelberg (where the Dolmetscher Institut has been in existence since 1930), Mainz (Germersheim), Montreal, Paris, Saarbrücken, Trieste, Vienna.[2] Since 1966 the School of Modern Languages at Bath University has offered a one-year postgraduate course for translators and interpreters. Courses are now also offered at a number of other institutions.

There is general agreement in Europe that the courses should be on the following lines:

1 native language
2 first and second foreign language
3 regional studies in the two foreign languages
4 special subject (not linguistic)
5 synchronic linguistics, including the theory of translation

This type of training no doubt corresponds to the demands made of its translators by German industry. They are selected according to criteria based on a translation test, professional experience, special subject knowledge, academic qualifications and a general cultural background.

It is preferable, however, to view translation as part of the process of information transfer in science and technology. A model for this is provided by courses suggested in *Science information personnel*.[3] which is produced under the auspices of the Modern Language Association of America and the National Science Foundation. The authors propose a new graduate school of science information and a one-year graduate curriculum leading to an MSC in information. This course is designed to qualify the student for one of the following vocations: translator, bibliographer, science information specialist, indexer, terminologist, technical editor, language specialist for information systems or acquisitions specialist.

The course in Language and Science Information is a theoretical study of the place of language in science information activities and a practical study of scientific information in foreign languages. It covers the following topics: relation of linguistics and semantics to the theory and structure of information systems; scientific nomenclature and terminology; standardization of language; theories of meaning; compilation of thesauri; machine translation; vocabulary of indexes; scientific dictionaries; theory of translating and intensive practice in translating.

A training of this kind would equip a translator to play an effective part in a foreign-language information system, and would enhance the status of the technical translator as a science information specialist. This would in turn attract greater numbers of talented people to the profession.

The American course is weaker than the European course on the cultural side, but it must be remembered that it is a graduate course and presupposes a good linguistic and cultural background in the student.

Translators may be freelances, devoting all or part of their time to translation, and may sometimes be employed by a private translation agency. Alternatively, they may work for an organization (a private firm or institution) on a full-time basis. Freelances can generally produce work more quickly but are (or should be) limited in range of subject-matter, and the presentation of the work cannot be as high as that of work from agencies or organizations. The agencies command a wider range of translators (more languages, more subjects) and can offer better presentation, but their fees will be higher and they will need more time. The work of a reputable agency will, however, be of higher quality as regards accuracy and presentation. Research organizations are willing to do some work for outside clients under certain circumstances, but obviously they will not wish to compete with private agencies.

Fees for translation vary according to the languages. Languages are graded according to rarity and difficulty. The Institute of Linguistics in Britain divides languages into various groups: Group I consists of French, Italian and Spanish; Group II Afrikaans, Dutch, German, Latin, Portuguese and the Scandinavian languages; Group III of Albanian, Bulgarian, the Celtic languages, Czech, Polish, Romanian and Russian; Group IV of Arabic, Estonian, Finnish, Hebrew and Hungarian; Group V of Japanese, Chinese and Korean. Charges for translating into or out of languages in Group I will be the lowest.

There are translators' organizations throughout Western Europe, many of them affiliated to the International Federation of Translators (FIT). FIT's headquarters are in Paris and it publishes the quarterly journal *Babel*, with financial support from Unesco.

Notes

𝕾𝕾𝕾𝕾𝕾𝕾

CHAPTER 1

1 Price, D. J. de Solla, 'Science and technology: Distinctions and inter-relations', in *Sociology of Science* (Penguin Modern Sociology Readings) (Harmondsworth 1972), p. 171.

2 Price, D. J. de Solla, in *Communication in Science* (Ciba Foundation symposium) (London 1967), p. 133. 'The totality of publications in the small languages is rapidly becoming more than 50 per cent of the world total ... We are stuck with the publication of the primary literature in more and more languages with each language becoming less than an absolute majority.'

3 Gould, C. J. and Stern, B. T., 'Foreign technical literature: a problem of costs, coverage and comprehension', *Aslib Proceedings*, XXIII (1971), pp. 571–6.

4 Jantsch, E. A., *A Study of Information Problems in the Electrotechnical Sector* (Paris: OECD 1965), p. 36.

5 Hanson, C. W. and Phillips, M., *The Foreign Language Barrier in Science and Technology* (London 1962), cited in Passman, S., *Scientific and Technological Communication* (London 1969), p. 119.

6 *Scientific and Technical Translating* (Paris: Unesco 1957), p. 14.

7 Wood, D. N., 'Chemical literature and the foreign-language problem', *Chemistry in Britain*, 11 (1966), p. 346.

8 Fëdorov, A., *Vvedenie v Teoriju Perevoda* ('Introduction to the Theory of Translation') (1953), quoted in Brang, P., 'Das Problem des Über-setzens', *Sprachforum*, 1 (1955), pp. 124–34.

9 Ortega y Gasset, J., 'Glanz und Elend der Übersetzung' (transl. from Spanish), in *Das Problem des Übersetzens*, ed. Störig, H. J. (Stuttgart 1963), p. 234.

CHAPTER 2

1 Erben, J., *Deutsche Grammatik* (Frankfurt-am-Main 1969), p. 21.

CHAPTER 3

1 Weinreich, U., *Languages in Contact* (The Hague 1964), p. 5.

CHAPTER 4

1 Chomsky, N., *Aspects of the Theory of Syntax* (Cambridge, Mass. 1965), p. 21.
2 *Atomkernenergie* (January/February 1965), p. 65.
3 Stockwell, R. P., Bowen, J. D. and Martin, J. W., *The Grammatical Structures of English and Spanish* (Chicago 1965), ch. 11. The *Contrastive Structure Series*, published by the University of Chicago, to which this book belongs, is very useful, though not all the books are of the same quality. An earlier work on contrastive studies is Leisi (1961). This is a study of semantic contrasts between German and English and is highly recommended. On the question of types of equivalence, see Nida (1964), Catford (1965) (especially ch. 3) and Quine (1967), for the distinction he makes between 'radical translation' and translation by means of 'traditional equations' (ch. 2).

CHAPTER 5

1 Bar-Hillel, Y., *Language and Information* (Reading, Mass. 1964), p. 51.
2 For a general view of the comparison of languages, see Ellis, J., *Towards a General Comparative Linguistics* (The Hague 1966).

CHAPTER 6

1 Erben, J., *Abriss der deutschen Grammatik* (Berlin 1959), pp. 15, 68 and 103. For further reading on the topics of this and chs. 7 and 8, see the *Duden*, Palmer's study of the English verb, Chomsky's *Aspects* and Koschmieder (1963). It is impossible to list all the relevant works available in German and English.

CHAPTER 8

1 Duden, *Die Grammatik der deutschen Gegenwartssprache* (Mannheim 1966).

CHAPTER 9

1 *Webster's Seventh New Collegiate Dictionary* (Springfield, Mass. 1963).
2 Mounin, G., *Les Problèmes Théoriques de la Traduction* (Paris 1963), pp. 85–6.
3 Leisi, *op. cit.*, p. 28.
4 Öhmann, S., *Wortinhalt und Weltbild* (Stockholm 1951), p. 159.
5 Berlin, B. and Kay, P., *Basic Color Terms* (Berkeley, Calif. 1969).
6 Leisi, *op. cit.*, p. 36.

CHAPTER 10

1 The following is an illustration of the extent to which meaning depends on the attendant circumstances:

Now 'pure water' is a queer sort of concept. Transistor makers require ion-free 'conductivity water', but, of course, they wouldn't mind a bit if it contained a fantastically lethal one part in a billion of botulin toxin, or was, in a health-scientist's terms, 'crawling with cholera'. The health scientist considers water fine when it's saturated with calcium and magnesium and bicarbonate, has a strong odor of hydrogen sulphide, corrodes out any boiler it doesn't first clog up with scale, and has enough free chlorine in it to chew pits in stainless steel. Just so long as it doesn't contain any living organisms.

And a thirsty man, in Kipling's immortal words, doesn't really mind if it 'was crawling and it stunk', or has some sand and mud in it, and college kids have demonstrated that a goldfish or two won't render it undrinkable.

For photographic purposes, well-filtered water, freed of sand, mud and the like, and preferably free of calcium and magnesium salts – they tend to precipitate and form scums with the photographic materials – is fine. *Analog* (November 1964), p. 9.

In this chapter and the next two I have used material from my MA thesis, 'Service translation viewed as conceptual transfer: An analysis in linguistic terms' (1968), submitted to the University of the Witwatersrand.

2 Lyons, J., *Structural Semantics* (Oxford 1963), p. 42. Öhmann (1951, p. 51) uses the expression *Begriffsgemeinschaft* (conceptual community).

CHAPTER 11

1 Ischreyt, H., *Studien zum Verhältnis von Sprache und Technik* (Düsseldorf 1965), pp. 38ff.
2 Gillam, D. J., 'The translation of horological texts', *Babel,* XVII (1971), pp. 7–9.
3 Savory, T. H., *The Art of Translation* (London 1957), p. 147.

CHAPTER 12

1 Kapp, P. O., 'The logic and psychology of science', *British Journal for the Philosophy of Science,* XV (1964–5), pp. 333–41.
2 Handel, S., *A Dictionary of Electronics* (Harmondsworth 1962).
3 Wüster, E., *Internationale Sprachnormung in der Technik* (Bonn 1966). This is the most authoritative work on the subject of standardizing technical terminology. It has been translated into Russian but not, as far as I know, into any other language. The best work available in English is Ray (1963). In addition, many articles have been written on this topic in English and German.

4 Wahlgren, J. H., 'Linguistic analysis of Russian chemical terminology', in *Proceedings, 1961, International Conference on Machine Translation of Languages* (London 1962).

5 Bloomfield, L., 'Linguistic Aspects of Science', in *International Encyclopedia of Unified Science* (Chicago 1939), p. 260.

6 Kapp, *op. cit.*

7 Hutten. E, H., *The Language of Modern Physics* (London 1956), p. 74.

8 Moon, P. and Spencer, D. E., 'Modern terminology for physics', *American Journal of Physics,* XVI (1948), pp. 100–4.

9 *Ibid.*

10 Townley, K. A., 'Clarity in geological writing', in *Science, 121* (1955), pp. 535–7.

11 Lockwood, J. F., *Flour Milling* (Liverpool 1952).

CHAPTER 13

On transliteration and transcription, see Gilyarevsky, R. S., and Grivin, V. S., *Languages Identification Guide* (Moscow 1970). For translation procedures, consult Catford (1965) and Vinay and Darbelnet (1964). Catford's approach is academic, that of Vinay and Darbelnet more practical. All three writers deal with translation in general.

1 These terms follow Catford, the account of modulation leans on Vinay and Darbelnet.

CHAPTER 14

1 Kade, O., 'Übersetzungskategorien und Rationalisierung', *Fremdsprachen,* III (1967), pp. 163–70.

2 Van der Pol, B., 'An iterative translation test', in *Information Theory, 3rd Symposium,* ed. C. Cherry (1956).

3 Winthrop, H., 'A proposed model and procedure for studying message distortion in translation', *Linguistics,* XXII (1966), pp. 98–112.

4 Miller, G. A., 'Human memory and the storage of information', *IRE Transactions on Information Theory,* IT-2, 3 (1956), pp. 129–37.
See also Miller, G. A., *The Psychology of Communication* (Harmondsworth 1970), especially 'Information and Memory'.

5 Krollmann, R., Schuck, J. and Winkler, U., 'Herstellung textbezogener Fachwortlisten mit einem Digitalrechner – ein Verfahren der automatischen Übersetzungshilfe', *Beiträge zur Sprachkunde und Informationsverarbeitung,* V (1965), pp. 7–30.

CHAPTER 15

1 For further reading on lexicography, see (in English) Householder, F. W. and Saporta, S. (eds), *Problems in Lexicography* (Bloomington,

Indiana 1967), and Zgusta, L., *Manual of Lexicography* (The Hague 1971). There is a much greater volume of literature on lexicography and lexicology in French, German, Italian, Russian and Spanish than in English, but there are some signs of growing interest.

2 *Bibliography of Interlingual Scientific and Technical Dictionaries* (Paris: Unesco 1969).

CHAPTER 16

1 Locke, W. W. and Booth, A. D., *Machine Translation of Languages* (New York 1956).

A good introduction to this subject is Delavenay, E., *Introduction to Machine Translation* (New York 1960). The collection edited by Booth, *Machine Translation* (Amsterdam 1967) gives a picture of later developments.

For computer languages, see Higman, B., *A Comparative Study of Programming Languages* (London 1967) and Sanderson, P. C., *Computer Languages) A practical guide to the chief programming languages* (London 1970).

2 Weaver, W., 'Translation', in Locke and Booth, *op. cit.*

3 Booth, A. D., in *Electronic Engineer's Reference Book*, ed. L. E. C. Hughes (London 1959).

4 Wilss, W., 'Automatische Sprachübersetzung', in *Sprache im technischen Zeitalter*, II (Berlin 1964), pp. 856–7.

5 Bar-Hillel, Y., 'A demonstration of the nonfeasibility of fully automatic high quality translation', in *Language and Information* (1964).

6 Garvin, P., *On Machine Translation* (The Hague 1972), p. 10.

7 Bar-Hillel, Y., *op. cit.*

8 Krollmann, F., Schuck, H. and Winkler, U., *op. cit.*

CHAPTER 18

The 25 September 1970 issue of the *Times Literary Supplement* was devoted to translation and contained articles on the conditions in which translators worked in a number of countries. One of the articles dealt with technical translation. Regular information is published in the international translators' journal, *Babel*, and in the *Incorporated Linguist*, the organ of the Institute of Linguists in Britain. *Aslib Proceedings* and Aslib's *Technical Translations Bulletin* also deal regularly with problems of interest to technical translators.

1 Kolers, P. A., 'Interlingual facilitation of short-term memory', *Journal of verbal learning and verbal behaviour*, v (1966), pp. 314–19.

2 Wilss, W., 'The school of translating and interpreting of the University of Saarbrücken', *Linguistic Reporter*, XI: 4 (1969), pp. 1–4.

3 Cohan, L. and Craven, K., *Science Information Personnel* (New York 1961).

Bibliography

𑀫𑀫𑀫𑀫𑀫𑀫

Brookes, B. C., 'Communicating research results', *Aslib Proceedings*, 16 (1964), pp. 7–21

Brower, R. (ed.), *On Translation* (New York 1966)

Cary, E. and Jumpelt, R. W. (eds), *Quality In Translation* (Oxford 1963).

Catford, J. C., *A Linguistic Theory of Translation* (Oxford 1965)

Cherry, C., *On Human Communication* (Cambridge, Mass. 1957)

Citroen, I. J. (ed.), *Ten Years of Translation* (London 1967)

Coseriu, E., 'Bedeutung und Bezeichnung im Lichte der strukturellen Semantik', in *Sprachwissenschaft und Übersetzen*, eds. P. Hartmann and H. Vernay (Munich 1970)

Duden, *Die Grammatik der deutschen Gegenwartssprache* (Mannheim 1966)

Garvin, P. (ed.), *Natural Language and the Computer* (New York 1963)

Greenberg, J. H. (ed.), *Universals of Language* (Cambridge, Mass. 1963)

Grogan, D., *Science and Technology, An introduction to the literature* (London 1970)

Halliday, M. A. K., *Grammar, Society and the Noun* (London 1967)

Jakobson, R., 'On linguistic aspects of translation', in Brower, *On Translation*

Jumpelt, R. W., *Die Übersetzung naturwissenschaftlicher und technischer Literatur* (Berlin-Schöneberg 1961)

Koschmieder, E., *Beiträge zur allgemeinen Syntax* (Heidelberg 1963)

Kufner, H. L., *The Grammatical Structures of English and German: A contrastive sketch* (Chicago 1962)

Leisi, E., *Der Wortinhalt, Seine Struktur im Deutschen und Englischen* (Heidelberg 1961)

Lyons, J., *Introduction to Theoretical Linguistics* (Cambridge 1968)

National Research Council, *Language and machines: computers in translation and linguistics: a report* (Washington 1966)

Nida, E., *Towards a Science of Translating* (Leiden 1964)

Palmer, E. R., *A Linguistic Study of the English Verb* (London 1965)

Passman, S., *Scientific and Technological Communication* (London 1969)

Potter, S., *Modern Linguistics*, 2nd edn (London 1967)

BIBLIOGRAPHY

Price, D. J. de Solla, *Little Science, Big Science* (New York 1963)
Quine, W. van O., *Word and Object* (Cambridge, Mass. 1967)
Ray, P. S., *Language Standardization* (The Hague 1963)
Vinay, J.-P. and Darbelnet, J., *Stylistique comparée du Français et de l'Anglais: Méthode de traduction* (Paris, Montreal 1964)

Index

Index

𒀭𒀭𒀭𒀭𒀭𒀭